D1128761

Jerry B. Marion, Ph.D.

and

Ronald C. Davidson, Ph.D.

Department of Physics and Astronomy,
University of Maryland

MATHEMATICAL REVIEW FOR THE PHYSICAL SCIENCES

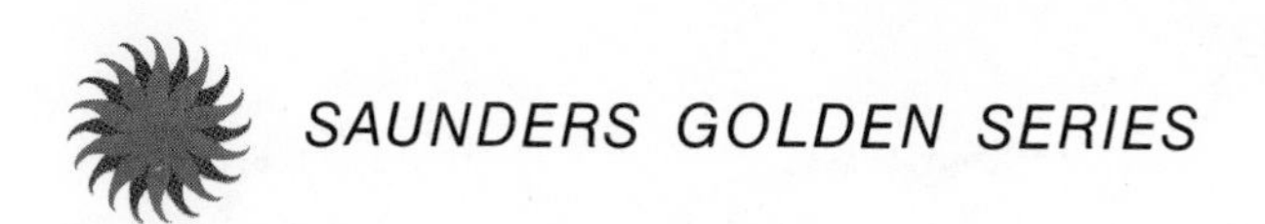

W. B. SAUNDERS COMPANY

Philadelphia, London, Toronto 1974

W. B. Saunders Company: West Washington Square
Philadelphia, Pa. 19105

12 Dyott Street
London, WC1A 1DB

833 Oxford Street
Toronto, Ontario M8Z 5T9, Canada

Mathematical Review for the Physical Sciences ISBN 0-7216-6076-2

Last digit is the print number: 9 8 7 6 5 4 3 2 1

PREFACE

Students who enter an introductory course in the physical sciences not having had a recent course requiring mathematical skills frequently find themselves unable to keep pace because of a lack of familiarity with the necessary mathematical tools. Such students may experience considerable difficulty in acquiring the requisite mathematical background because of a lack of appropriate refresher courses in most colleges and universities or because no suitable (and brief) review book can be found. This book has been designed to fill the need of these students. We have assembled here a short overview of all the various types of mathematical topics that can reasonably be expected to be necessary in an introductory-level course in the physical sciences. Because our intention is to assist the student in an *introductory* physical sciences course, we have confined our discussions to elementary mathematical ideas.

We have attempted to present mathematical ideas from the standpoint of the *scientist.* That is, we are concerned here only with the *use* of mathematical techniques, not with rigorous developments. Wherever possible we have introduced mathematical ideas with physical examples, and worked problems have frequently been included to emphasize each topic. In addition, there are numerous exercises which the student should use to test his understanding of each new idea and technique. Answers to these exercises will be found in the back of this book.

While using this book the student should keep in mind that this is not a physical sciences text—it is a mathematical aid and should be used in addition to a serious study of the assigned text.

JERRY B. MARION
RONALD C. DAVIDSON

CONTENTS

CHAPTER ONE

DEALING WITH NUMBERS AND UNITS

Man began to make real progress in understanding the physical world around him only when he started to think in *quantitative* terms. The results of measurements and observations expressed in *numbers* are the keys that permit us to unlock the secrets of Nature and to understand and appreciate the orderliness of natural phenomena. Therefore, we begin this survey of essential mathematics by examining the way we use numbers as descriptive tools in the physical sciences.

1.1 OPERATIONS WITH NUMBERS

There are two classes of numbers that we will be using:

Rational Numbers. These are the positive and negative *integers* and *zero*,

$$\cdots, -3, -2, -1, 0, 1, 2, 3, \cdots$$

and the fractions m/n, where m and n are integers.

Irrational Numbers. These are numbers expressible as continuing decimals. For example, $\sqrt{2} = 1.41421 \cdots$, where the dots mean that the series of decimal numbers continues without end. Also, $\pi = 3.14159 \cdots$. (Ratios of integers are *not* irrational numbers even though they may be expressed as continuing decimals; for example, $2/3 = 0.66666 \cdots$ is a *rational* number).

The operations of *addition* (denoted by +) and *multiplication* (denoted by × or ·) have the property that the result is independent of the *order* of the factors. That is,

$$4 + 3 = 7 \qquad 4 \times 3 = 12$$

$$3 + 4 = 7 \qquad 3 \times 4 = 12$$

In general, we can express this property in the following way:

$$a + b = b + a$$
$$a \times b = b \times a \qquad (1.1)$$
for any numbers a and b

The operations of addition and multiplication are said to be *commutative*. Notce that *subtraction* (denoted by $-$) and *division* (denoted by $\div$) are *not* commutative operations. For example,

$$4 - 3 \neq 3 - 4; \quad 4 \div 3 \neq 3 \div 4$$

(The symbol $\neq$ means "is *not* equal to".)

The operation of subtraction is exemplified by the following:

$$11 - 6 = 5$$
$$-11 + 6 = -5$$

The rule for subtracting a *negative* number is to *change the sign and add.* Thus,

$$15 - (-7) = 15 + 7 = 22$$

In general, we can write

$$a - (-b) = a + b \qquad (1.2)$$

When multiplication involves negative numbers, we have the following rules:

$$(-a) \times (b) = -(a \times b)$$
$$(a) \times (-b) = -(a \times b) \qquad (1.3)$$
$$(-a) \times (-b) = +(a \times b)$$

The operation of *division* involves taking the *ratio* of two numbers. For example,

$$1 \div 2 = 1/2 = 0.50$$
$$2 \div 3 = 2/3 = 0.6666 \cdots$$

In general,

$$a \div b = \frac{a}{b} \qquad (1.4)$$

The sign rules for division involving negative numbers are similar to those for multiplication (Eq. 1.3):

$$(-a) \div (b) = -\frac{a}{b}$$

$$(a) \div (-b) = -\frac{a}{b} \qquad (1.5)$$

$$(-a) \div (-b) = +\frac{a}{b}$$

EXERCISES

1. $18 + 15 =$ ____ (Ans. 1)*
2. $18 - 15 =$ ____ (Ans. 2)
3. $-18 + 15 =$ ____ (Ans. 3)
4. $18 - (-15)$ ____ (Ans. 4)
5. $18 \times 15 =$ ____ (Ans. 5)
6. $(-18) \times 15 =$ ____ (Ans. 6)
7. $18 \times (-15) =$ ____ (Ans. 7)
8. $(-18) \times (-15) =$ ____ (Ans. 8)
9. $18 \div 15 =$ ____ (Ans. 9)
10. $(-18) \div 15 =$ ____ (Ans. 10)
11. $18 \div (-15) =$ ____ (Ans. 11)
12. $(-18) \div (-15) =$ ____ (Ans. 12)

1.2 FRACTIONS

In the fraction a/b, we call a the *numerator* and b the *denominator*. In many computations, it is useful to replace fractions by their *decimal equivalents*. For example,

$$\frac{4}{10} = 0.40$$

$$\frac{5}{8} = 0.625$$

---

*These answers will be found beginning on page 105.

A fraction can be manipulated in a variety of ways without changing its value. For example, we can *multiply* the numerator and denominator by the same number and not change the value of the fraction. Thus,

$$\frac{2}{3} = \frac{2 \times 2}{3 \times 2} = \frac{4}{6}$$

$$\frac{2}{3} = \frac{2 \times 13}{3 \times 13} = \frac{26}{39}$$

In general,

$$\frac{a}{b} = \frac{a \times c}{b \times c} \tag{1.6}$$

where a, b, and c are any three numbers.

We can also *divide* the numerator and denominator of a fraction by the same number and not change the value of the fraction. For example,

$$\frac{20}{30} = \frac{20 \div 10}{30 \div 10} = \frac{2}{3}$$

or,

$$\frac{20}{30} = \frac{20 \div 2}{30 \div 2} = \frac{10}{15}$$

In general,

$$\frac{a}{b} = \frac{a \div c}{b \div c} \tag{1.7}$$

or, equivalently,

$$\frac{a}{b} = \frac{a/c}{b/c} \tag{1.8}$$

This rule is frequently used when we simplify fractions by cancelling the same quantity from both the numerator and denominator. For example, $(3a/3b)$ can be expressed as (a/b) because both the numerator and the denominator can be divided by 3. We say that we have *cancelled* 3 from both the numerator and the denominator. Cancellation is usually denoted by a slash (/) through the numbers in question:

$$\frac{3a}{3b} = \frac{\cancel{3}a}{\cancel{3}b} = \frac{a}{b}$$

Cancellation can be used to simplify fractions if there is a factor common to both the numerator and the denominator. For example,

$$\frac{144}{156} = \frac{\cancel{12} \times 12}{\cancel{12} \times 13} = \frac{12}{13} = 0.923 \cdots$$

$$\frac{81}{27} = \frac{\cancel{27} \times 3}{\cancel{27} \times 1} = \frac{3}{1} = 3$$

Sometimes the denominator of a fraction is itself a fraction. For example, consider the quantity

$$\frac{3}{\frac{2}{3}}$$

To simplify this fraction, we multiply both numerator and denominator by a number that reduces the denominator to 1. Here, the appropriate number is the fraction 3/2:

$$\frac{3 \times \frac{3}{2}}{\frac{2}{3} \times \frac{3}{2}} = \frac{3 \times \frac{3}{2}}{1} = \frac{9}{2}$$

In general, fractions of this type can be simplified according to the rule

$$\frac{a}{\frac{b}{c}} = a \times \frac{c}{b} \tag{1.9}$$

That is, we *invert the denominator and multiply.* For example,

$$\frac{7}{\frac{4}{5}} = 7 \times \frac{5}{4} = \frac{35}{4} = 8.75$$

$$\frac{\frac{1}{3}}{\frac{7}{4}} = \frac{1}{3} \times \frac{4}{7} = \frac{4}{21} = 0.190 \cdots$$

We frequently need to add (or subtract) fractions. One possible procedure is to bring the fractions to a *common denominator* and then perform the required operations. For example,

$$\begin{aligned}\frac{3}{5}+\frac{1}{6}&=\frac{3\times 6}{5\times 6}+\frac{1\times 5}{6\times 5}\\&=\frac{18}{30}+\frac{5}{30}\\&=\frac{18+5}{30}\\&=\frac{23}{30}\\&=0.766\cdots\end{aligned}$$

Note that $30=5\times 6=6\times 5$ is the *common denominator* in the above example. As a further example,

$$\begin{aligned}\frac{1}{8}-\frac{3}{16}&=\frac{1\times 2}{8\times 2}-\frac{3}{16}\\&=\frac{2}{16}-\frac{3}{16}\\&=\frac{2-3}{16}\\&=-\frac{1}{16}\\&=-0.0625\end{aligned}$$

Another procedure for adding (or subtracting) fractions is to write each fraction as its *decimal equivalent* and then add (or subtract). For example,

$$\begin{aligned}\frac{1}{8}-\frac{3}{16}&=0.1250-0.1875\\&=-0.0625\end{aligned}$$

which is the same result as that obtained by performing the subtraction with fractions.

Exercises

Express the following as decimal equivalents:

1. $\frac{111}{333}$ (Ans. 13)

2. $\frac{14}{56}$ (Ans. 14)

3. $\frac{12}{36}$ (Ans. 15)

4. $\frac{9}{3/7}$ (Ans. 16)

5. $\frac{16}{9}$ (Ans. 17)

6. $\frac{7}{2/5}$ (Ans. 18)

7. $\frac{3}{1/8}$ (Ans. 19)

8. $\frac{1}{5}+\frac{3}{7}$ (Ans. 20)

9. $\frac{2}{3}-\frac{5}{9}$ (Ans. 21)

10. $3\frac{9}{10}$ (Ans. 22)

11. $1\frac{1}{3}+6\frac{2}{3}$ (Ans. 23)

12. $\frac{15}{8}+\frac{1}{2}$ (Ans. 24)

(Hint: whenever possible, simplify the fraction *before* converting to decimal notation).

1.3 POWERS OF TEN

On the basis of certain astronomical measurements, we know that the distance from the Sun to the star Alpha Centauri is 4 070 000 000 000 000 000 centimeters. And we know that the mass of a single atom of hydrogen is 0.000 000 000 000 000 000 000 001 67 grams. This method of specifying the distance or the mass with a large number of zeroes is awkward and cumbersome. To overcome this difficulty in the expression of large or small numbers, we use a compact notation employing "powers of ten."

Multiplying 10 by itself a number of times, we find

$$10 \times 10 = 100 = 10^2$$

$$10 \times 10 \times 10 = 1000 = 10^3$$

$$10 \times 10 \times 10 \times 10 \times 10 = 100\,000 = 10^5$$

The number of times that 10 is multiplied together (that is, the number of *zeroes* that appear in the answer) is used in the result as a *superscript* to the 10. This superscript is called the *exponent* of 10 or the *power* to which 10 is raised. Also,

$$10^1 = 10$$

and, by convention,

$$10^0 = 1$$

If we express two numbers as powers of ten and then multiply these numbers, we have

$$10^2 \times 10^3 = (10 \times 10) \times (10 \times 10 \times 10) = 10^5 = 10^{(2+3)}$$

The general rule for this kind of operation is

$$10^n \times 10^m = 10^{(n+m)} \tag{1.10}$$

Just as 10^3 means $10 \times 10 \times 10$, we can raise *any* number to a power n by multiplying that number by itself n times. Thus,

$$2^3 = 2 \times 2 \times 2 = 8$$
$$4^2 = 4 \times 4 = 16$$
$$(1.5)^3 = 1.5 \times 1.5 \times 1.5 = 3.375$$

If a power of 10 appears in the denominator of an expression, the exponent is given a negative sign:

$$\frac{1}{10} = 0.1 = 10^{-1}$$

$$\frac{1}{1000} = 0.001 = \frac{1}{10^3} = 10^{-3}$$

In general,

$$\frac{1}{10^m} = 10^{-m} \tag{1.11}$$

Multiplying a *positive* power of 10 by a *negative* power of 10 gives

$$10\,000 \times 0.01 = 10^4 \times \frac{1}{10^2} = 10^4 \times 10^{-2} = 10^{(4-2)} = 10^2$$

$$\frac{100}{1000} = \frac{10^2}{10^3} = 10^{(2-3)} = 10^{-1} = 0.1$$

In general, combining the rules given in Equations 1.10 and 1.11, we have

$$\frac{10^n}{10^m} = 10^n \times 10^{-m} = 10^{(n-m)} \tag{1.12}$$

Exercises

1. $10^9 = \_\_\_$ (Ans. 25)
2. $10\,000\,000 = \_\_\_$ (Ans. 26)
3. $(10^3) \times (10^5) = \_\_\_$ (Ans. 27)
4. $3^4 = \_\_\_$ (Ans. 28)
5. $(1.2)^3 = \_\_\_$ (Ans. 29)
6. $(1000)^4 = \_\_\_$ (Ans. 30)
7. $\frac{1}{100\,000} = \_\_\_$ (Ans. 31)
8. $0.000\ 001 = \_\_\_$ (Ans. 32)
9. $\frac{10^6}{1000} = \_\_\_$ (Ans. 33)
10. $\frac{0.001}{10^4} = \_\_\_$ (Ans. 34)
11. $\frac{10^3 \times 0.1}{10\,000 \times 10^2} = \_\_\_$ (Ans. 35)
12. $\frac{4^3 \times 3^4}{2^4 \times 3^3} = \_\_\_$ (Ans. 36)

1.4 CALCULATIONS WITH POWERS OF TEN

By using powers of 10, many types of calculations are made considerably easier. First, we note that *any* number can be expressed in terms of a power of 10 by writing, for example,

$$6400 = 6.4 \times 10^3$$

$$0.0137 = 1.37 \times 10^{-2}$$

$$970\,000 = 0.97 \times 10^6$$

Notice that in changing from an ordinary number to a number expressed as a power of 10, the exponent of 10 corresponds to the number of places that the decimal has been moved. (A positive exponent means that the decimal has been moved to the left and a negative exponent means that the decimal has been moved to the right.) Thus, in expressing 6400 in terms of a power of 10, we move the decimal three places to the left to obtain 6.4×10^3. And in expressing 0.0137 in terms of a power of 10, we move the decimal two places to the right, obtaining 1.37×10^{-2}.

Usually, we express a quantity in this notation by writing the multiplying factor as a number between 0.1 and 10. That is, we write $240\ 000 = 2.4 \times 10^5$ instead of 240×10^3.

$$\frac{42\ 000\ 000}{3000} = \frac{4.2 \times 10^7}{3 \times 10^3} = \frac{4.2}{3} \times \frac{10^7}{10^3} = 1.4 \times 10^4$$

$$0.0012 \times 0.000\ 003 = (1.2 \times 10^{-3}) \times (3 \times 10^{-6})$$
$$= (1.2 \times 3) \times (10^{-3} \times 10^{-6}) = 3.6 \times 10^{-9}$$

Exercises

1. Express the distance from the Sun to Alpha Centauri (in cm) as a power of 10. (See Section 1.3.) (Ans. 37)
2. Express the mass of a hydrogen atom (in grams) as a power of 10. (See Section 1.3.) (Ans. 38)
3. $\frac{6\ 400\ 000}{1600} =$ ____ (Ans. 39)
4. $3200 \times 0.0004 =$ ____ (Ans. 40)
5. $\frac{24\ 000}{0.012} =$ ____ (Ans. 41)
6. $\frac{160 \times 0.0024}{0.32 \times 480\ 000} =$ ____ (Ans. 42)
7. $16\ 000 \times 0.03 \times 0.12 =$ ____ (Ans. 43)

1.5 PREFIXES

When discussing physical quantities, it frequently proves convenient to use a *prefix* to a unit instead of a power of 10. For example, *centi-* means $\frac{1}{100}$, so *centi*-meter or centimeter (cm) means $\frac{1}{100}$ of a meter (m):

$$1\ \text{m} = 100\ \text{cm} = 10^2\ \text{cm};\ \ 1\ \text{cm} = 0.01\ \text{m} = 10^{-2}\ \text{m}$$

Similarly, *milli-* means $\frac{1}{1000} = 10^{-3}$ and *mega-* means 10^6:

$$1\ \text{m} = 10^3\ \text{millimeters (mm)};\ \ 1\ \text{mm} = 10^{-3}\ \text{m}$$

$$\$1\ 000\ 000 = 1\ \text{megabuck}$$

Table 1.1 lists some of the most frequently used prefixes.

TABLE 1.1 PREFIXES EQUIVALENT TO POWERS OF 10

Prefix	Symbol	Power of 10
giga-	G	10^9 *
mega-	M	10^6 *
kilo-	k	10^3
centi-	c	10^{-2}
milli-	m	10^{-3}
micro-	μ	10^{-6}
nano-	n	10^{-9}

*$10^6 = 1$ *million*. In the U.S., $10^9 = 1$ *billion*, but the European convention is that $10^9 = 1000$ million and that 1 billion $= 10^{12}$; the prefix *giga-* is internationally agreed on to represent 10^9.

EXERCISES

1. 1 kilometer (km) = ____ meters (m) (Ans. 44)
2. 1 nanosecond (ns) = ____ seconds (s) (Ans. 45)
3. 1 mm = ____ km (Ans. 46)
4. (1 km) × (1 m) = ____ cm^2 (Ans. 47)

1.6 SQUARE ROOTS

When the number 3 is multiplied by itself, the result is 9: $3 \times 3 = 9$. But 3×3 is the *square* of 3, so we can also write $3^2 = 9$. We say that 9 is the *square* of 3 and that 3 is the *square root* of 9. We indicate a square root with the symbol $\sqrt{}$. Thus $\sqrt{9} = 3$. Also,

$$4 \times 4 = 16; \text{ therefore, } \sqrt{16} = 4$$

$$17 \times 17 = 289; \text{ therefore, } \sqrt{289} = 17.$$

We sometimes need to evaluate square roots of the form $\sqrt{1-(0.8)^2}$. The procedure is first to evaluate the quantity under the square root sign, and then to take the square root. Since

$$1 - (0.8)^2 = 1 - 0.64 = 0.36$$

we find

$$\sqrt{1-(0.8)^2} = \sqrt{0.36}$$

TABLE 1.2 SOME USEFUL SQUARE ROOTS

$\sqrt{2} = 1.414$	$\sqrt{25} = 5$
$\sqrt{3} = 1.732$	$\sqrt{36} = 6$
$\sqrt{4} = 2$	$\sqrt{49} = 7$
$\sqrt{5} = 2.236$	$\sqrt{64} = 8$
$\sqrt{9} = 3$	$\sqrt{81} = 9$
$\sqrt{16} = 4$	$\sqrt{100} = 10$

Because

$$(0.6) \times (0.6) = 0.36$$

we conclude that

$$\sqrt{1 - (0.8)^2} = 0.6$$

Similarly,

$$\sqrt{1 - 0.9375} = \sqrt{0.0625} = 0.25$$

The last equality follows because

$$(0.25) \times (0.25) = 0.0625$$

The square roots shown in Table 1.2 occur so frequently that it is worthwhile to remember these. A complete list of square roots for the integers 1 to 100 is given on page 100.

EXERCISES

1. $\sqrt{16} =$ ____ (Ans. 48)
2. $\sqrt{81} =$ ____ (Ans. 49)
3. $\sqrt{36} =$ ____ (Ans. 50)
4. $\sqrt{0.25} =$ ____ (Ans. 51)
5. $\sqrt{17} =$ ____ (Ans. 52)
6. $\sqrt{1 - (0.6)^2} =$ ____ (Ans. 53)
7. $\sqrt{\dfrac{32 \times 10^5}{2^3 \times 10^3}} =$ ____ (Ans. 54)
8. $\sqrt{93} =$ ____ (Ans. 55)

(*Hint*: The table of square roots on page 100 may be useful).

1.7 MATHEMATICAL NOTATION

In ordinary equations we use the symbol = to denote equality of two quantities:

$$y = 16.27\ t^2 \quad \text{or} \quad A \times B = C$$

Even if we did not know the factor 16.27 which occurs in the above equation, we could still state that y is *proportional to* t^2, and we would write

$$y = kt^2 \quad \text{or} \quad y \propto t^2$$

Or, if we knew only that the factor is *approximately* equal to 16, we would write

$$y \cong 16\, t^2$$

The symbols $<$ and $>$ mean, respectively, *is less than* and *is greater than*; for example,

area of Canada $>$ area of Argentina

mass of the Earth $<$ mass of Jupiter

If a quantity is *very much smaller* or *very much larger* than another quantity, we use a double symbol:

area of Canada $>>$ area of Luxembourg

mass of the Earth $<<$ mass of the Milky Way Galaxy

We frequently find it convenient to use a shorthand notation to indicate the *change* in the value of a quantity. If an object is located at the position $x_1 = 2$ cm at a certain time and if at a later time the location is $x_2 = 9$ cm, we say that the distance moved (or the change in x) is $x_2 - x_1 = 9\text{ cm} - 2\text{ cm} = 7$ cm. That is, we take the *final* position (x_2) and subtract from it the *initial* position (x_1). This change in x is often denoted by the symbol Δx:

$$\Delta x = x_{\text{final}} - x_{\text{initial}} = x_2 - x_1 \tag{1.13}$$

The symbol Δx does *not* imply the product of Δ and x, but means "the change in x" or "an increment of x." In general, a Greek delta, Δ, in front of a quantity means the *change* in that quantity; e.g., $t_2 - t_1 = \Delta t =$ time difference. Δt can be either *positive* ($\Delta t > 0$) or *negative* ($\Delta t < 0$), depending on whether t_2 is greater or smaller than t_1.

The meanings of the symbols we will find useful in this book are summarized in Table 1.3.

TABLE 1.3 MATHEMATICAL SYMBOLS AND THEIR MEANINGS

Symbol	Meaning
$=$	is equal to
$\propto$	is proportional to
$\cong$	is approximately equal to
$>$ ($<$)	is greater (less) than
$\gg$ ($\ll$)	is much greater (less) than
Δx	change in x

Exercises

Insert the appropriate symbol between the following pairs of quantities (in some cases there may be two appropriate symbols—give both):

1. height of Mont Blanc Mt. Everest (Ans. 56)
2. area of Canada area of Brazil (Ans. 57)
3. mass of an apple mass of an orange (Ans. 58)
4. If $\frac{a}{b}=6$, then a b (Ans. 59)

1.8 FUNDAMENTAL UNITS OF MEASURE

Physical quantities not only have *magnitudes*, specified by *numbers*, but also *dimensions* or *units*. It makes no sense to say that a certain length is "75" unless we also state the appropriate units—feet, meters, miles, or whatever. Although we encounter a wide variety of physical quantities that require units for their complete specification—for example, distance, force, energy, momentum, and electric field strength—these various units can be expressed in terms of only *three* fundamental quantities. The basic units of physical measure are those of *length, mass,* and *time*—the dimensions of all other physical quantities can be expressed in terms of the units of these three. For example, *speed* is measured by the distance traveled in a certain time and so the dimensions of speed are *length/time.*

The system of units most widely used in the physical sciences is the meter-kilogram-second or MKS system. Also in use at the present time

TABLE 1.4 FUNDAMENTAL UNITS OF MEASURE

	Metric		
	MKS	*CGS*	**British**
Length	Meter (m) 1 m = 100 cm	Centimeter (cm)	Yard (yd) 1 yd = 36 in 1 in = 2.54 cm
Mass	Kilogram (kg) 1 kg = 1000 g	Gram (g)	Pound-mass (lb) 1 lb = 453.59 g
Time	Second (s)	Second (s)	Second (s)

is the *British engineering system,* in which the standard units are the yard, the pound-mass, and the second. It seems inevitable that the metric system will replace the British system and will become the international standard system within a matter of years.

The fundamental units of measure in the metric and British systems are summarized in Table 1.4.

If we wish to convert a certain length from one system of units to another, we use the following procedure. Since 1 in = 2.54 cm, we can form the ratios

$$\frac{2.54 \text{ cm}}{1 \text{ in}} = 1, \quad \text{or} \quad \frac{1 \text{ in}}{2.54 \text{ cm}} = 1$$

These ratios are just *unity,* so we can multiply any quantity by either ratio without changing the value. Thus,

$$15 \text{ in} = (15 \text{ in}) \times \left(\frac{2.54 \text{ cm}}{1 \text{ in}}\right) = 38.10 \frac{\text{cm-in}}{\text{in}} = 38.10 \text{ cm}$$

where the *inches* in numerator and denominator cancel. Also,

$$48 \text{ cm} = (48 \text{ cm}) \times \left(\frac{1 \text{ in}}{2.54 \text{ cm}}\right) = 18.9 \text{ in}$$

$$36 \text{ kg-m} = (36 \text{ kg-m}) \times \left(\frac{10^3 \text{ g}}{1 \text{ kg}}\right) \times \left(\frac{10^2 \text{ cm}}{1 \text{ m}}\right) = 3.6 \times 10^6 \text{ g-cm}$$

$$60 \text{ mi/hr} = (60 \text{ mi/hr}) \times \left(\frac{5280 \text{ ft}}{1 \text{ mi}}\right) \times \left(\frac{1 \text{ hr}}{3600 \text{ s}}\right) = 88 \text{ ft/s}$$

Note the cancellation of units in each of these examples.

Conversion factors for units of length are summarized in Table 1.5, and conversion factors for units of mass are summarized in Table 1.6.

TABLE 1.5 CONVERSION FACTORS FOR UNITS OF LENGTH

	m	cm	yd	in
1 m =	1	100	1.094	39.37
1 cm =	0.01	1	0.01094	0.3937
1 yd =	0.9144	91.44	1	36
1 in =	0.0254	2.54	1/36	1

1 in = 2.54 cm (exactly)

1 mi = 1760 yd = 5280 ft (statute mile)

= 1.609 km = 1.609×10^3 m = 1.609×10^5 cm

1 km = 0.6214 mi

Areas and volumes:

1 in^2 = 6.452 cm^2

1 ft^2 = 929 cm^2

1 in^3 = 16.39 cm^3

1 ft^3 = 2.832×10^4 cm^3

Any quantity which does not have units is said to be *dimensionless* or a *pure number.* For example, the ratio of two physical quantities with the same dimensions is dimensionless:

$$R = \frac{8 \text{ cm}}{4 \text{ cm}} = 2$$

The quantity R is a *pure number.*

TABLE 1.6 CONVERSION FACTORS FOR UNITS OF MASS

	kg	g	lb
1 kg =	1	10^3	2.205
1 g =	10^{-3}	1	2.205×10^{-3}
1 lb =	0.4536	453.6	1

1 lb = 453.59237 g (exactly)

Example 1.8.1

One acre consists of 43 560 ft^2. Express this figure in m^2. We first note from Table 1.5 that

$$1 \text{ ft}^2 = 929 \text{ cm}^2$$

Furthermore,

$$1 \text{ m}^2 = (100 \text{ cm}) \times (100 \text{ cm}) = 10^4 \text{ cm}^2$$

Therefore,

$$1 \text{ acre} = (4.356 \times 10^4 \text{ ft}^2) \times \left(\frac{929 \text{ cm}^2}{1 \text{ ft}^2}\right) \times \left(\frac{1 \text{ m}^2}{10^4 \text{ cm}^2}\right)$$
$$= 4047 \text{ m}^2$$

Note the cancellation of units in this example.

EXERCISES

1. 1 km = ____ cm (Ans. 60)
2. 3 lb = ____ kg (Ans. 61)
3. 43 mm = ____ in (Ans. 62)
4. 1 km = ____ mi (Ans. 63)
5. 7.2 yd/s = ____ m/s (Ans. 64)
6. 14 m/s = ____ km/hr (Ans. 65)
7. 107 lb-in = ____ g-cm (Ans. 66)
8. 1 year = ____ s (Ans. 67)
9. Express 1 mi^2 in cm^2. (Ans. 68)
10. 1 gallon = 231 in^3. Express this figure in cm^3. (Ans. 69)
11. How many acres are there in 1 mi^2? (Refer to Example 1.8.1) (Ans. 70)
12. One metric ton is equal to 10^3 kg. Express this figure in lb. (Ans. 71)

CHAPTER TWO

ALGEBRA

2.1 PROPERTIES OF EQUATIONS

An *equation* expresses the *equality* between two quantities or combinations of quantities. For example,

$$2 + 4 = 3 + 3$$

expresses the equality between $2 + 4$ (which is equal to 6) and $3 + 3$ (which is also equal to 6). Similarly,

$$A + B = C + D$$

expresses the equality between $A + B$ and $C + D$.

Quantities that are equal to the same quantity are also equal to one another. Suppose that we are given

$$A + B = F$$

and

$$C + D = F$$

Therefore, since $A + B$ and $C + D$ are both equal to F, they are equal to one another:

$$A + B = C + D$$

Equations can be manipulated in various ways without affecting the validity of the equality. We summarize the various operations that can be applied to the equation $A + B = C + D$:

(a) The same quantity can be *added* to both sides of an equation without affecting the equality:

$$(A + B) + M = (C + D) + M \tag{2.1}$$

(b) The same quantity can be *subtracted* from both sides of an equation without affecting the equality:

$$(A+B)-M=(C+D)-M \tag{2.2}$$

(c) Both sides of an equation can be *multiplied* by the same quantity without affecting the equality:

$$(A+B)\times M=(C+D)\times M \tag{2.3}$$

(d) Both sides of an equation can be *divided* by the same quantity without affecting the equality:

$$\frac{(A+B)}{M}=\frac{(C+D)}{M} \tag{2.4}$$

(e) Both sides of an equation can be *raised to the same power* without affecting the equality:

$$(A+B)^n=(C+D)^n \tag{2.5}$$

(f) *Taking the square root* of both sides of an equation does not affect the equality:

$$\sqrt{(A+B)}=\sqrt{(C+D)} \tag{2.6}$$

In general, if the same mathematical operation is applied to both sides of an equation, the result is a valid equation.

Another important procedure which we will often use is *factoring*. When a certain number is common to two terms in a sum, the number can be written as a common multiplying factor. For example,

$$3\times 5+3\times 7=3\times(5+7)$$

or,

$$6ab+9ac=3a\times(2b+3c)$$

The general rule is

$$(AC+BC)=(A+B)\times C \tag{2.7}$$

This is called the *distributive law* of multiplication.

2.2 LINEAR EQUATIONS

Usually we want to solve an equation for an unknown quantity. For example, if we are given an equation such as

$$3x-7=0$$

we wish to determine the value of the unknown quantity x.

We will consider in this section equations which can be reduced to this form

$$ax + b = 0 \tag{2.8}$$

where a and b are specified numbers and x is the unknown quantity. This is a *linear* equation, that is, an equation for the unknown quantity x in which x appears only to the *first* power. (Equations in which x^2 appears are *quadratic* equations and will be discussed in Section 2.3.)

The basic operations described in Section 2.1 are sufficient to solve all linear equations. The object is to select the proper operations to isolate the unknown quantity x.

Example 2.2.1

Solve $3x - 7 = 0$.
Step 1. Add 7 to both sides:

$$3x - \not{7} + \not{7} = 0 + 7$$

$$3x = 7$$

Step 2. Divide both sides by 3:

$$\frac{\not{3}x}{\not{3}} = \frac{7}{3}$$

$$x = \frac{7}{3}$$

Example 2.2.2

Solve $3x + 4 = x - 1$

Step 1. Subtract $(x + 4)$ from both sides:

$$(3x + 4) - (x + 4) = (x - 1) - (x + 4)$$

$$3x - x + \not{4} - \not{4} = \not{x} - \not{x} - 1 - 4$$

$$2x = -5$$

Step 2. Divide both sides by 2:

$$\frac{\not{2}x}{\not{2}} = \frac{-5}{2}$$

$$x = -\frac{5}{2}$$

Example 2.2.3

Solve $\frac{2}{3} = \frac{5}{x}$

Step 1. Multiply both sides by x:

$$\frac{2x}{3} = \frac{5\cancel{x}}{\cancel{x}}$$

$$\frac{2x}{3} = 5$$

Step 2. Multiply both sides by 3:

$$\cancel{3} \times \frac{2x}{\cancel{3}} = 3 \times 5$$

$$2x = 15$$

Step 3. Divide both sides by 2:

$$\frac{\cancel{2}x}{\cancel{2}} = \frac{15}{2}$$

$$x = \frac{15}{2}$$

Example 2.2.4

An automobile travels at a constant speed of 40 miles per hour. How much time is required to go 340 miles?

The basic equation is

$$\text{distance} = \text{velocity} \times \text{time}$$

In symbols this becomes

$$x = vt \tag{1}$$

The unknown quantity is the time t, and solving for t,

$$t = \frac{x}{v} = \frac{340 \text{ mi}}{40 \text{ mi/hr}} = 8.5 \text{ hr} \tag{2}$$

Notice the way in which the units are manipulated:

$$\frac{\text{mi}}{\frac{\text{mi}}{\text{hr}}} = \frac{\cancel{\text{mi}}}{\frac{\cancel{\text{mi}}}{\cancel{\text{hr}}}} \times \frac{\text{hr}}{\cancel{\text{hr}}} = \text{hr} \tag{3}$$

EXERCISES

In each case, solve for x:

1. $2x+1=0$ (Ans. 72)
2. $4x+5=3$ (Ans. 73)
3. $\frac{4}{3}x=7$ (Ans. 74)
4. $3x+8=50-4x$ (Ans. 75)
5. $12x-a=b+7+4x$ (Ans. 76)
6. $\frac{3}{2}x+12=\frac{7}{2}x-4$ (Ans. 77)
7. $\frac{a}{x+b}=c$ (Ans. 78)
8. $\frac{3}{x-4}=\frac{7}{x-5}$ (Ans. 79)

2.3 QUADRATIC EQUATIONS

Sometimes we encounter an equation such as

$$x^2=a$$

Again, we wish to find the value of x. According to Equation 2.6, it is permissible to take the *square root* of both sides of the equation:

$$\sqrt{x^2}=\sqrt{a}$$
$$x=\sqrt{a}$$

If we *square* this result, we find the original equation, $x^2=a$. But notice that if we square $x=-\sqrt{a}$, we also obtain $x^2=a$, because $(-\sqrt{a})\cdot(-\sqrt{a}) = a$. Therefore, when we take the square root of the equation $x^2=a$, we do not determine the *sign* of x. We must write the result as

$$x=\pm\sqrt{a}$$

where the symbol $\pm$ means that $\sqrt{a}$ carries either a positive sign *or* a negative sign. Usually, in a physical problem, the situation will determine which sign is appropriate. For example, if we are calculating the *mass* of some object, the negative sign has no meaning (there are no *negative* masses), and the positive sign is to be used.

Example 2.3.1.

Solve

$$5x^2 - 6 = x^2 + 30 \tag{1}$$

Step 1. Subtract $x^2 - 6$ from both sides:

$$(5x^2 - 6) - (x^2 - 6) = (x^2 + 30) - (x^2 - 6)$$

$$(5 - 1)\,x^2 - \cancel{6} + \cancel{6} = (\cancel{1} - \cancel{1})\,x^2 + (30 + 6) \tag{2}$$

$$4x^2 = 36$$

Step 2. Divide both sides by 4:

$$\frac{\cancel{4}x^2}{\cancel{4}} = \frac{36}{4}$$

$$x^2 = 9$$

Step 3. Take the square root of both sides:

$$x = \pm\sqrt{9}$$

$$x = \pm 3$$

Example 2.3.2

If an object is dropped from a certain height, the distance x (measured in feet) that it falls downward during a time t (measured in seconds) is given by

$$x = 16t^2$$

Suppose that a ball is dropped from the top of a 100-ft building. When will it strike the ground?

In this problem, we are required to find the value of t when $x = 100$ ft. We write

$$100 = 16t^2$$

Dividing by 16, we have

$$t^2 = \frac{100}{16}$$

Instead of dividing 100 by 16 and then taking the square root to obtain t, we can simplify the calculation by noting that $\sqrt{100} = \pm 10$ and $\sqrt{16} = \pm 4$. Therefore,

$$t = \pm \frac{\sqrt{100}}{\sqrt{16}} = \pm \frac{10}{4} = \pm 2.5 \text{ s}$$

The time of fall must, of course, be positive, so the answer is $t = 2.5$ s. In physical problems, the *positive* square root is usually (but not always) the desired root.

When we take the square root of a quantity, the sign of the quantity under the radical must be *positive*. That is, we can write $\sqrt{16} = 4$, because $4 \times 4 = 16$. But we cannot express $\sqrt{-16}$ in terms of real numbers because there is no real number that multiplied by itself yields -16. (We can express quantities such as $\sqrt{-16}$ in terms of *imaginary* numbers, but we will have no need for such numbers here.)

Quadratic equations always involve the second power (or the *square*) of the unknown quantity and can be reduced to the general form,

$$a x^2 + b x + c = 0 \tag{2.9}$$

The solution to this equation is

$$x = \frac{-b \pm \sqrt{b^2 - 4\,ac}}{2\,a} \tag{2.10}$$

In order for the solution to be real, the quantity under the radical must be positive (or zero). That is, we must have $b^2 > 4\,ac$ or $b^2 = 4\,ac$. Notice that when $b^2 > 4\,ac$, the $\pm$ sign between the two terms means that there are *two* solutions. Which is the desired solution must be determined by examining the physical situation; sometimes both solutions have physical meaning.

Example 2.3.3

Solve the equation

$$x^2 + 4x + 4 = 0 \tag{1}$$

We identify

$$a = 1, \; b = 4, \; c = 4 \tag{2}$$

and we note that $b^2 = 4ac$. Then, using Equation 2.10, we have

$$x = \frac{-4 \pm \sqrt{4^2 - 4 \cdot 1 \cdot 4}}{2 \times 1} = \frac{-4 \pm \sqrt{16 - 16}}{2} \tag{3}$$

Since the square root term is zero, we find

$$x = -2 \tag{4}$$

The answer is readily verified to be correct by substitution of (4) into (1).

Example 2.3.4

Solve the equation

$$7x^2 - 8x + 1 = 0 \tag{1}$$

We identify

$$a = 7, \; b = -8, \; c = 1 \tag{2}$$

and we see that $b^2 = 64$ and $4ac = 28$, so that $b^2 > 4ac$. Using Equation 2.10, the solution is

$$x = \frac{-(-8) \pm \sqrt{(-8)^2 - 4 \cdot 7 \cdot 1}}{2 \times 7}$$

$$= \frac{8 \pm \sqrt{64 - 28}}{14} = \frac{8 \pm \sqrt{36}}{14}$$

$$= \frac{8 \pm 6}{14} \tag{3}$$

There are now two possible solutions, one corresponding to each of the signs of the square-root term. We label these x_+ and x_-:

$$x_+ = \frac{8 + 6}{14} = 1 \tag{4}$$

$$x_- = \frac{8 - 6}{14} = \frac{2}{14} = \frac{1}{7} \tag{5}$$

Again, it is easy to verify by substitution into (1) that each solution is valid.

Example 2.3.5

In your study of the motion of objects, you will learn that at a time t, the position x of an object moving in either direction along a straight line is given by

$$x = x_0 + v_0 t + \frac{1}{2} a t^2 \tag{1}$$

where x_0 is the position at time $t = 0$ and v_0 is the velocity at $t = 0$; the quantity a is the acceleration of the object. In the metric system, the units of v_0 are m/s and the units of a are (m/s)/s or m/s² (read: "meters per second per second"). If an object falls freely toward the Earth, the acceleration is 9.8 m/s² (near the surface of the Earth).

Suppose that a ball is thrown directly downward with a velocity of 7 m/s from a height of 20 m. When will the ball strike the ground?

In order to use (1), we must decide on the values of the various quantities, x, x_0, v_0, and a. If we call the moment of release of the ball $t = 0$, then $v_0 = 7$ m/s. We also know that $a = 9.8$ m/s². The values of x and x_0 depend on where we elect to place the position we call $x = 0$. Let us take the *downward* direction to be the positive direction (so that the acceleration is positive) and let us call the position of release $x = 0$. Then, $x_0 = 0$ and the final position (the surface of the Earth) is $x = 20$ m.

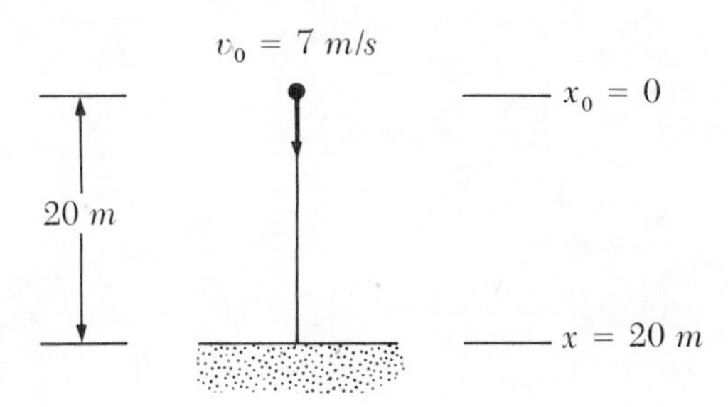

Equation (1) now becomes (we do not write the units of the various quantities because we know they are all metric units)

$$20 = 0 + 7t + \frac{1}{2} \times 9.8 \times t^2$$

Rearranging, we have

$$4.9\,t^2 + 7t - 20 = 0 \tag{2}$$

Comparing (2) with Equation 2.9, we see that

$$a = 4.9,\ b = 7,\ c = -20 \tag{3}$$

Inserting these numbers into Equation 2.10, we find

$$t = \frac{-7 \pm \sqrt{(7)^2 - (4)(4.9)(-20)}}{9.8}$$

$$= \frac{-7 \pm \sqrt{49 + 392}}{9.8}$$

$$= \frac{-7 \pm \sqrt{441}}{9.8}$$

$$= \frac{-7 \pm 21}{9.8}$$

Therefore, the two solutions are

$$t_+ = \frac{-7 + 21}{9.8} = \frac{14}{9.8} = 1.43 \text{ s}$$

$$t_- = \frac{-7 - 21}{9.8} = -\frac{28}{9.8} = -2.86 \text{ s}$$

Clearly, t_+ is the desired solution.

EXERCISES

Solve the following equations for x:

1. $2x^2 - 450 = 0$ (Ans. 80)
2. $3x^2 - 300 = 0$ (Ans. 81)
3. $6x^2 - 7 = 5x^2 + 50$ (Ans. 82)
4. $\frac{2}{x^2} = \frac{1}{50}$ (Ans. 83)
5. $3x^2 - 6x - 9 = 0$ (Ans. 84)
6. $x^2 - 9x + 20 = 0$ (Ans. 85)
7. $3x^2 + 8x - \frac{19}{4} = 0$ (Ans. 86)
8. $\frac{x^2}{3} - 300 = 0$ (Ans. 87)

2.4 AVERAGE VALUES

The results of quantitative measurements in the physical sciences are expressed in terms of numbers, and often the *average value* (or *mean value*) of a set of similar measurements is a useful concept. To evaluate the *average value* of a collection of N numbers, we *add the individual numbers in the collection and then divide by* N.

As an example, a spectroscopist makes *four* measurements of the wavelength λ (Greed *lambda*) of light emitted by a yellow light source. Suppose that the results of his measurements are:

$$\lambda_1 = 5.26 \times 10^{-7}\ \text{m}$$

$$\lambda_2 = 5.21 \times 10^{-7}\ \text{m}$$

$$\lambda_3 = 5.25 \times 10^{-7}\ \text{m}$$

$$\lambda_4 = 5.24 \times 10^{-7}\ \text{m}$$

Here the subscripts, 1, 2, etc., attached to λ label the particular wavelength measurement (first, second, etc.). The difference in the measured values of wavelength may result from variations in the light source and/or errors inherent in the method of measurement. From the definition of *average value* we find that the *average wavelength* inferred from the four measurements ($N=4$) is

$$\begin{aligned} &\text{Average wavelength} \\ &= \frac{1}{4}(\lambda_1 + \lambda_2 + \lambda_3 + \lambda_4) \\ &= \frac{1}{4}(5.26 \times 10^{-7}\text{m} + 5.21 \times 10^{-7}\text{m} + 5.25 \times 10^{-7}\text{m} + 5.24 \times 10^{-7}\text{m}) \\ &= \frac{1}{4} \times (20.96 \times 10^{-7}\text{m}) \\ &= 5.24 \times 10^{-7}\text{m} \end{aligned}$$

The above result can also be stated as

$$\bar{\lambda} = 5.24 \times 10^{-7}\text{m}$$

where $\bar{\lambda}$ denotes the *average wavelength* (a *super bar* is often used to signify *average value*).

It is important that all quantities have the *same* units before calculating the mean value. If some of the above wavelength measurements had been expressed in centimeters (cm) and the rest in meters (m), it would have been necessary to convert all measurements to the same units (cm *or* m) before calculating $\bar{\lambda}$.

As a further example, the individual heights (h) in a family of three are determined to be

$$h_1 = 42\ \text{in}$$

$$h_2 = 5\ \text{ft}\ 6\ \text{in}$$

$$h_3 = 6\ \text{ft}$$

To calculate the average height ($\bar{h}$) we first convert all values of h to the same units. Because 12 in = 1 ft, we find

$$h_1 = 3.5 \text{ ft}$$
$$h_2 = 5.5 \text{ ft}$$
$$h_3 = 6 \text{ ft}$$

From the definition of average value, the average height is

$$\bar{h} = \frac{1}{3}(h_1 + h_2 + h_3)$$
$$= \frac{1}{3}(3.5 \text{ ft} + 5.5 \text{ ft} + 6 \text{ ft})$$
$$= \frac{1}{3}(15 \text{ ft})$$

or

$$\bar{h} = 5 \text{ ft}$$

Note that $N = 3$ in this example.

Exercises

Calculate mean values for the following arrays:

1. 1, 2, 3, 4, 5 (Ans. 88)
2. 0, 1, 2, 3, 4, 5 (Ans. 89)
3. 6 g, 10 g, 8 g (Ans. 90)
4. 40 s, 1 min, 4 min, 140 s (Ans. 91)
5. 100 m, 1 km (Ans. 92)

CHAPTER THREE

GEOMETRY

3.1 CIRCLES AND ANGULAR MEASURE

One of the most common geometric forms that we encounter in the physical sciences is the *circle*. Every circle is characterized by a *radius*, which is the straight-line distance from the center to any point on the circle (Fig. 3.1). The *diameter* of a circle is defined to be twice the radius or the distance from one side of the circle to the other through the center: $d = 2r$. The distance completely around a circle (called the *circumference*) is related to the diameter by a special irrational number designated by the symbol π (Greek *pi*):

$$\text{Circumference} = \pi \times \text{diameter}$$

or,

$$c = \pi d = 2\pi r \tag{3.1}$$

In ancient times there was no known method by which the precise value of π could be determined. Often, the approximate value, $\pi \cong \frac{22}{7} = 3\frac{1}{7}$, was used, and for some calculations this number is sufficiently close to the actual value:

$$\pi = 3.14159 \cdots \tag{3.2}$$

FIGURE 3.1 Parts of a circle.

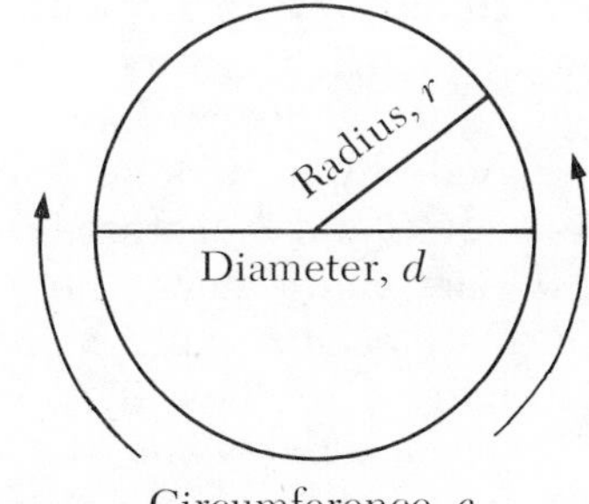

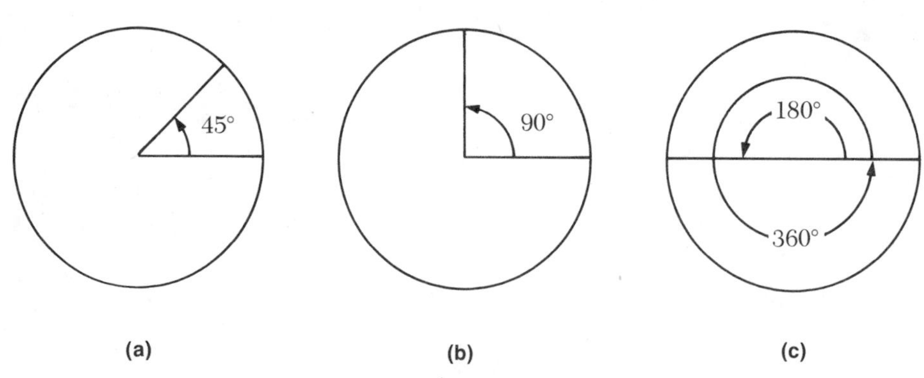

FIGURE 3.2 Various angles and their relation to a circle.

If we draw two different radii in a particular circle, as in Figure 3.2, we can specify the *angle* between the radii in terms of *degrees of arc* (or, simply, *degrees*), the symbol for which is °. A complete circle corresponds to 360°. Figure 3.2*a* shows two radii with an included angle of 45°. In Figure 3.2*b* the angle between the radii is 90°, and the two lines are said to be at *right angles* or *perpendicular*. If the two radii form a straight line (a diameter), as in Figure 3.2*c*, the angle is 180°. Finally, if we continue the angle completely around the circle, the angle is 360° (Fig. 3.2*c*).

In order to specify angles that are not exactly equal to some number of degrees, we divide the degree into smaller parts called *minutes of arc* (or *arc min*) and *seconds of arc* (or *arc sec*):

$$1° = 60 \text{ arc min} = 60'$$

$$1' = 60 \text{ arc sec} = 60''$$

Therefore, $1° = 3600$ arc sec $= 3600''$.

Sometimes we use decimal fractions of a degree instead of arc min and arc sec. For example,

$$17°\ 30' = 17.5°$$

$$32°\ 14'\ 37'' = 32.2436°$$

In Section 3.5 we will discuss a way to determine the value of an angle by using the methods of trigonometry. But if the two lines that define the angle are drawn on a piece of paper, we can always measure the angle by using a device called a *protractor*. Figure 3.3*a* shows two lines *AOC* and *BOD* that cross. Four angles are formed: angles *AOD* and *BOC* are *acute* angles (less than 90°) and are equal; angles *AOB* and *DOC* are *obtuse* angles (greater than 90°) and are also equal. As shown in Figure 3.3*b*, these angles can be measured with a protractor by placing the base line along *AOC* with the center hole at *O*. The angle *BOC* is

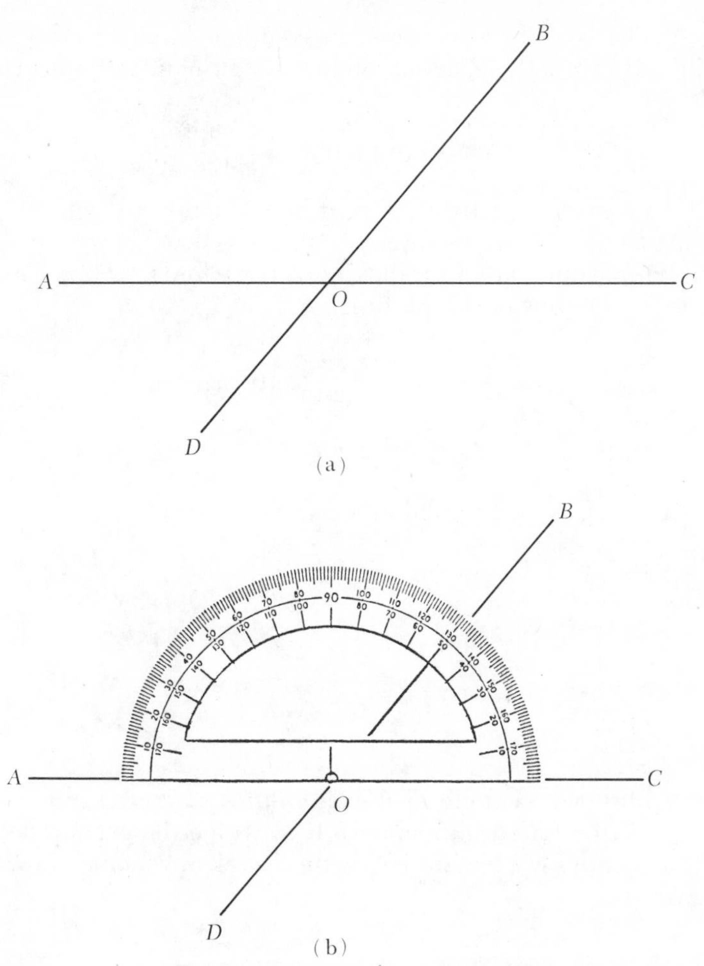

FIGURE 3.3 (a) The lines *AOC* and *BOD* form angles which are measured by the protractor in (b).

read from the inner scale as 50°. Notice that the outer scale gives the angle *AOB* as 130°. (We know that angle *AOB* must be 130° because *AOC* is a straight line and corresponds to 180°; therefore angle *AOB* is $180° - 50° = 130°$.)

Example 3.1.1

The nautical mile was originally defined to be the distance corresponding to 1 minute of arc of longitude at the Earth's equator. Since 1959 the nautical mile has been defined to be 1852 m. Use this information to compute the radius of the Earth in km.

The number of arc min in a complete circle is $360 \times 60 = 21\,600$. Therefore, the circumference in nautical miles is

$$c = (21\,600 \text{ arc min}) \times \left(\frac{1 \text{ naut mi}}{1 \text{ arc min}}\right)$$
$$= 2.16 \times 10^4 \text{ naut mi}$$

Solving Equation 3.1 for the radius r, we find $r = c/2\pi$. Therefore, the radius of the Earth is

$$r = \frac{c}{2\pi} = \frac{1}{2\pi} \times (2.16 \times 10^4 \text{ naut mi})$$
$$= 3.44 \times 10^3 \text{ naut mi}$$

Finally, converting to kilometers, we have

$$r = (3.44 \times 10^3 \text{ naut mi}) \times \left(\frac{1852 \text{ m}}{1 \text{ naut mi}}\right) \times \left(\frac{1 \text{ km}}{10^3 \text{ m}}\right)$$
$$= 6.38 \times 10^3 \text{ km}$$

An additional example of the properties of a circle in a scientific problem will be found in Section 6.1, in which the ancient determination of the Earth's circumference by the Greek astronomer Eratosthenes is discussed.

EXERCISES

1. What is the circumference of a circle that has a radius of 3.3 m? (Ans. 93)
2. What is the diameter of a circle that has a circumference of 3.3 m? (Ans. 94)
3. Express 13.75° in terms of degrees and minutes of arc. (Ans. 95)
4. Express 3.6° in terms of minutes of arc. (Ans. 96)
5. Suppose that you are viewing the full Moon. Imagine that two lines are drawn from your eye to the opposite ends of a diameter of the Moon's disk. What do you think would be the angle between the two lines:
(a) 5°, (b) 0.5°, (c) 5′, (d) 5″? (Ans. 97)

3.2 PROPERTIES OF SIMPLE GEOMETRIC FORMS

In many situations in the physical sciences we encounter geometrical constructions involving *triangles.* We therefore need to know some of the properties of these figures:

(a) The sum of the interior angles of any triangle is 180°. Figure 3.4 shows three cases in which this statement can be tested.

(b) Any triangle that has two sides of equal length is called an *isosceles* triangle; if all three sides are of equal length, the triangle is *equilateral* (see Fig. 3.5).

(c) If two triangles have two sides and the included angle (the angle between these sides) equal, then the triangles are equivalent and are said to be *congruent.* Similarly, if the triangles have two angles and the included side equal, they are also congruent (see Fig. 3.6). If two triangles have been proven to be congruent, then each pair of corresponding sides and each pair of corresponding angles are equal. For example, in the upper pair of triangles in Figure 3.6 the sides opposite the angles θ are equal.

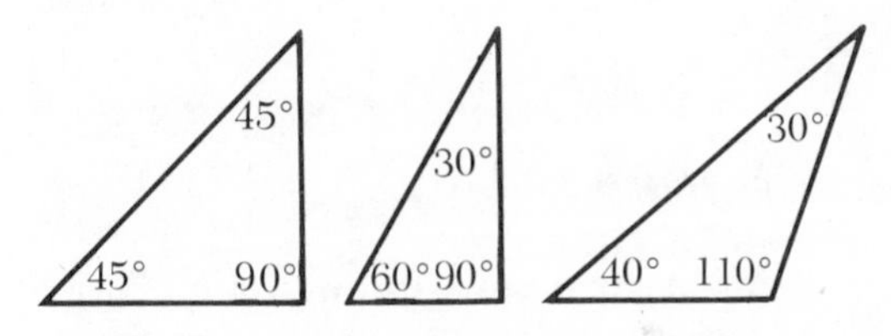

FIGURE 3.4 The interior angles of any triangle sum to 180°.

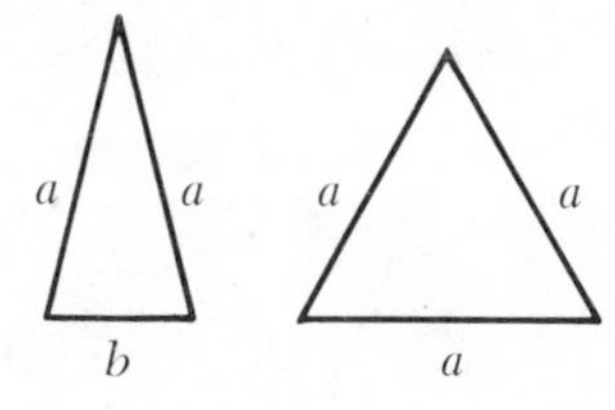

FIGURE 3.5 An *isosceles* and an *equilateral* triangle.

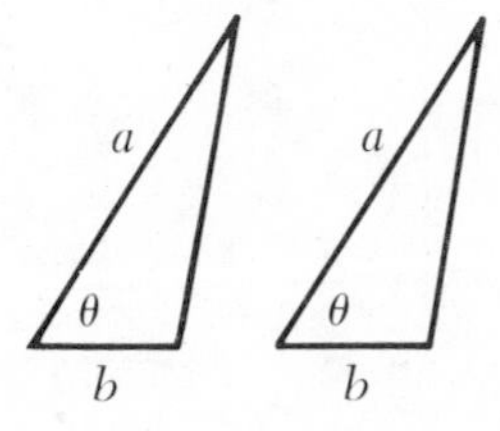

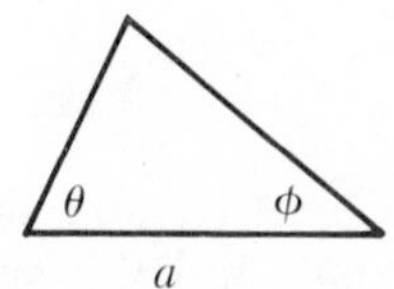

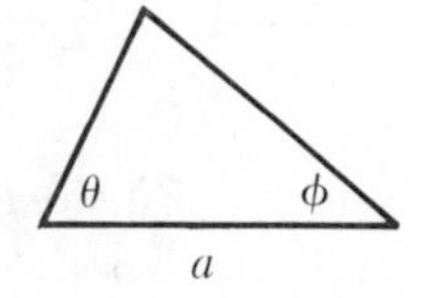

FIGURE 3.6 Two pairs of *congruent* triangles.

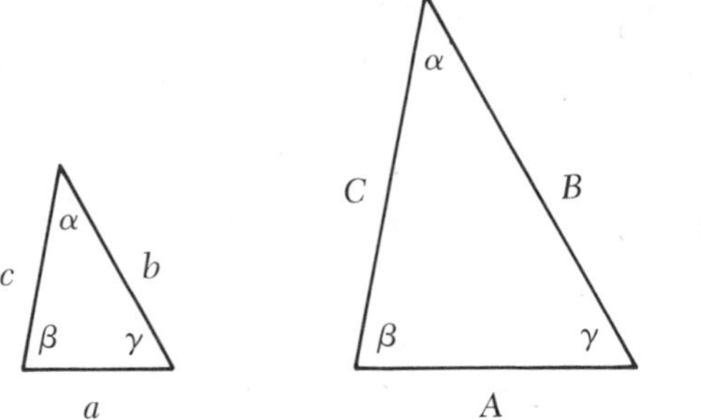

FIGURE 3.7 Two similar triangles.

(d) Two triangles are said to be *similar* if each has identical interior angles, as shown in Figure 3.7. Similar triangles have the same *shape*, but they need not be congruent. Similar triangles have the important property that the ratios of their sides are equal; thus, for the triangles in Figure 3.7, we have

$$\frac{a}{b} = \frac{A}{B}; \quad \frac{a}{c} = \frac{A}{C}; \quad \frac{a}{A} = \frac{c}{C}; \text{ etc.} \tag{3.3}$$

Example 3.2.1

A road runs through a rectangular field as shown in the diagram below. A surveyor measures the remaining length of the sides of the field and finds: $AB = 150$ m, $BC = 300$ m, and $DE = 110$ m. What will the surveyor find for the side EF?

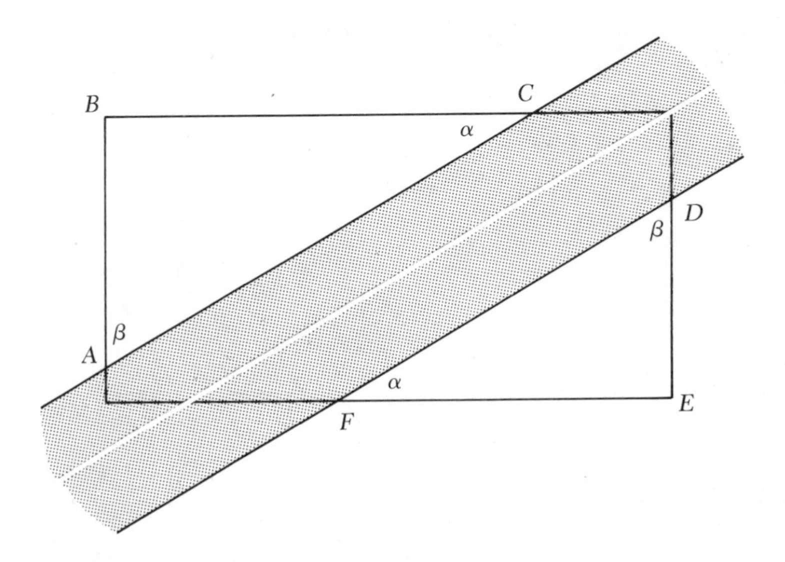

Because the sides of the road are parallel, the angles ACB and DFE are equal, and the angles BAC and FDE are also equal. The corners, B and E, are both formed by right angles (90°). All three interior angles of the triangles ABC and DEF are equal, and so the triangles are *similar*. Therefore, we can write an equality between the ratios,

$$\frac{AB}{DE} = \frac{BC}{EF}$$

Solving this equation for the unknown side EF and supplying the measured values, we find

$$EF = BC \times \frac{DE}{AB}$$

$$= (300 \text{ m}) \times \frac{(110 \text{ m})}{(150 \text{ m})} = 220 \text{ m}$$

In addition to triangles, we often deal with other geometric shapes such as rectangles, parallelepipeds, spheres, and cylinders. The most important properties of these figures that we need to know are the areas and, in the case of solid figures, the volumes. These properties are summarized in Table 3.1 on following page.

EXERCISES

1. What area of the field in Example 3.2.1 remains after the road is cut through? (Ans. 98)
2. What is the volume of a cube with sides of length 5 cm? (Ans. 99)
3. A cube has a surface area of 24 cm^2. What is its volume? (Ans. 100)
4. What is the area of a circle whose diameter is 8 cm? (Ans. 101)
5. What is the volume of a sphere whose radius is 3 cm? (Ans. 102)
6. A cylinder is 4 cm high and has a radius of 2 cm. What is the volume? (Ans. 103)
7. The radius of the Moon is 1.74×10^6 m. What is the volume of the Moon? (Ans. 104)

3.3 DENSITY

The fundamental quantities length, time, and mass can be combined in various ways to provide units for different physical quantities. For

TABLE 3.1 PROPERTIES OF SOME SIMPLE GEOMETRIC FORMS

Form	Property
Rectangle:	Area $= ab$
Parallelepiped:	Volume $= abc$
Right Triangle:	Area $= \frac{1}{2} ab$
Circle:	Circumference $= 2\pi r$ Area $= \pi r^2$ Diameter $= d = 2r$
Sphere:	Surface area $= 4\pi r^2$ Volume $= \frac{4}{3} \pi r^3$
Cylinder :	Surface area $= 2\pi r^2 + 2\pi rh$ $= 2\pi r\,(r + h)$ Volume $= \pi r^2 h$

example, *speed* or *velocity* is measured in terms of *length per unit time* (miles per hour, meters per second, or some other combination). In the physical sciences, we encounter many of these *derived* quantities. In addition to velocity, we use acceleration, force, momentum, work, energy, power, and several others. Even though we attach special names to the units for many of these quantities, it should be remembered that the *fundamental* definition of any physical quantity can always be made in terms of length, time, and mass.

As an example of a derived quantity, let us consider a simple but important case—*density*. If we cut a bar of iron into a number of pieces with various sizes and shapes, the pieces will all have different masses. But each piece still consists of *iron* and it must have some property that is characteristic of iron. If one of the pieces is twice as large as another, the mass must also be twice as great. A piece three times as large would have three times the mass, and so forth. That is, the ratio of the mass to the volume is constant for a particular substance—this ratio is called the *density*:

$$\frac{\text{Mass}}{\text{Volume}} = \text{Density}$$

or, in symbols,

$$\frac{M}{V} = \rho \tag{3.4}$$

Mass is measured in kilograms (or grams) and volume is measured in cubic meters (or cubic centimeters). Therefore, the units of density are kg/m³ or g/cm³. Some representative densities are listed in Table 3.2. Notice that the density of water is 1.00 g/cm³. In fact, the kilogram was originally defined as the mass of 1000 cm³ of water.

Example 3.3.1

Lead bricks (used for shielding radioactive materials) commonly have dimensions of 2 in × 4 in × 8 in. What is the mass of such a brick?

$$\begin{aligned} \text{Volume} = V &= (2\text{ in}) \times (4\text{ in}) \times (8\text{ in}) = 64\text{ in}^3 \\ &= (64\text{ in}^3) \times \left(\frac{2.54\text{ cm}}{1\text{ in}}\right)^3 \\ &= (64\text{ in}^3) \times \left(\frac{16.39\text{ cm}^3}{1\text{ in}^3}\right) \\ &= 1049\text{ cm}^3 \end{aligned} \tag{1}$$

TABLE 3.2 DENSITIES OF SOME MATERIALS

Material	Density	
	g/cm³	kg/m³
Gold	19.3	1.93×10^4
Mercury	13.6	1.36×10^4
Lead	11.3	1.13×10^4
Iron	7.86	7.86×10^3
Aluminum	2.70	2.70×10^3
Water	1.00	1.00×10^3
Air	0.0013	1.3

Using $\rho = 11.3$ g/cm³ from Table 3.2, we have

$$M = \rho V = \left(\frac{11.3\ \text{g}}{\text{cm}^3}\right) \times (1049\ \text{cm}^3)$$
$$= 1.185 \times 10^4\ \text{g} \tag{2}$$
$$= 11.85\ \text{kg}$$

or, in British units,

$$M = (1.185 \times 10^4\ \text{g}) \times \left(\frac{1\ \text{lb}}{453.59\ \text{g}}\right)$$
$$= 26.12\ \text{lb}$$

Example 3.3.2

What is the average density of the Earth?
The radius of the Earth is $R_E = 6.38 \times 10^6$ m, so that

$$V = \frac{4}{3}\pi R_E^3 = \frac{4}{3}\pi(6.38 \times 10^6\ \text{m})^3$$
$$= 1.09 \times 10^{21}\ \text{m}^3$$

The mass of the Earth is $M_E = 5.98 \times 10^{24}$ kg; thus,

$$\rho = \frac{M_E}{V} = \frac{5.98 \times 10^{24}\ \text{kg}}{1.09 \times 10^{21}\ \text{m}^3} = 5.49 \times 10^3\ \text{kg/m}^3$$
$$= 5.49\ \text{g/cm}^3$$

This is the density averaged throughout the entire Earth. Actually, the core has a considerably higher density (about 12 g/cm³) and the mantle (the rocky material near the surface) has a density of about 3 g/cm³.

EXERCISES

1. What is the mass of a cube of iron with sides of length 2 cm? (Ans. 105)
2. What is the mass of 20 m³ of air at normal conditions? (Ans. 106)
3. A sphere of aluminum has a radius of 10 cm. What is the mass? (Ans. 107)
4. A pipe with a diameter of 10 cm contains a column of water 10 m high. What is the mass of the water? (Ans. 108)

5. What is the difference in mass between two spheres (each with radius = 5 cm), one made of lead and the other made of gold? (Ans. 109)

6. What is the density (the average density) of the Sun? (Radius $= 6.96 \times 10^8$ m, mass $= 1.99 \times 10^{30}$ kg.) How does the density of the Sun compare with that of water? (Ans. 110)

3.4 THE DISTANCE BETWEEN TWO POINTS

One of the important results of plane geometry is the *Pythagorean theorem*, which states that the square of the length of the side opposite the right angle (called the *hypotenuse*) of a right triangle is equal to the sum of the squares of the lengths of the other two sides. That is (refer to Fig. 3.8),

$$c^2 = a^2 + b^2 \text{ (right triangle)} \tag{3.5}$$

Thus, if we know the lengths of any two sides of a right triangle, we can always find the length of the third side. If, in Figure 3.8, we know $c = 14$ cm and $a = 5$ cm, then

$$\begin{aligned} b &= \sqrt{c^2 - a^2} = \sqrt{(14)^2 - (5)^2} \text{ cm} = \sqrt{196 - 25} \text{ cm} \\ &= \sqrt{171} \text{ cm} = 13.08 \text{ cm} \end{aligned}$$

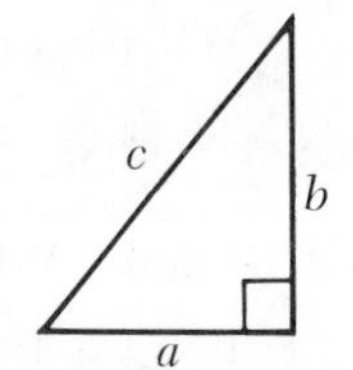

FIGURE 3.8 Identification of sides in a right triangle.

Example 3.4.1

A motorist travels east from a starting point P at a constant speed of 50 mi/hr for 3 hr. He then turns north and for 2 hr he maintains a constant speed of 65 mi/hr. At the end of the 5-hr trip, how far is he from P? That is, what is the distance from P to Q?

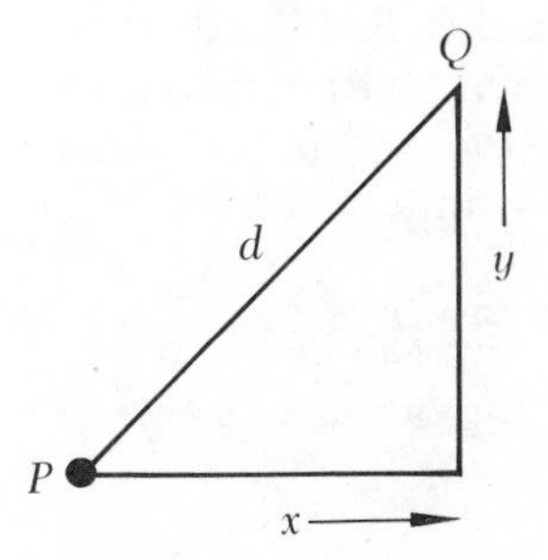

The first leg of the trip is

$$x = (\text{speed}) \cdot (\text{time})$$
$$= vt = (50 \text{ mi/hr}) \cdot (3 \text{ hr}) = 150 \text{ mi}$$

and the second leg is

$$y = (65 \text{ mi/hr}) \cdot (2 \text{ hr}) = 130 \text{ mi}$$

Therefore,

$$d = \sqrt{x^2 + y^2} = \sqrt{(150)^2 + (130)^2} \text{ mi}$$
$$= \sqrt{22500 + 16900} \text{ mi} = \sqrt{39400} \text{ mi}$$
$$= \sqrt{3.94} \cdot 10^2 \text{ mi} = 198.5 \text{ mi}$$

EXERCISES

1. Construct a triangle with sides equal to 3 cm, 4 cm, and 5 cm. Is this a right triangle? How can you tell? (Ans. 111)
2. The hypotenuse of a right triangle is 13 m in length and the shortest side is 5 m in length. What is the length of the third side? (Ans. 112)
3. The hypotenuse of a right triangle has a length of 20 cm, and the other two sides are equal in length. What are the lengths of these sides? (Ans. 113)
4. Point A lies due north of point B at a distance of 20 m. Point C lies due east of point B at a distance of 40 m. How far is it from point A to point C? (Ans. 114)

3.5 SOME SIMPLE ASPECTS OF TRIGONOMETRY

Consider a right triangle as shown in Figure 3.9. The angle ACB is a right angle and the angles BAC and ABC are labeled θ and ϕ, respectively. The ratio of the length of the side opposite θ to the length of the hypotenuse (that is, side AB) is called the *sine* of the angle θ. This quantity is abbreviated as sin θ. Therefore,

$$\sin \theta = \frac{\text{side opposite angle } \theta}{\text{hypotenuse}} = \frac{a}{c} \qquad (3.6)$$

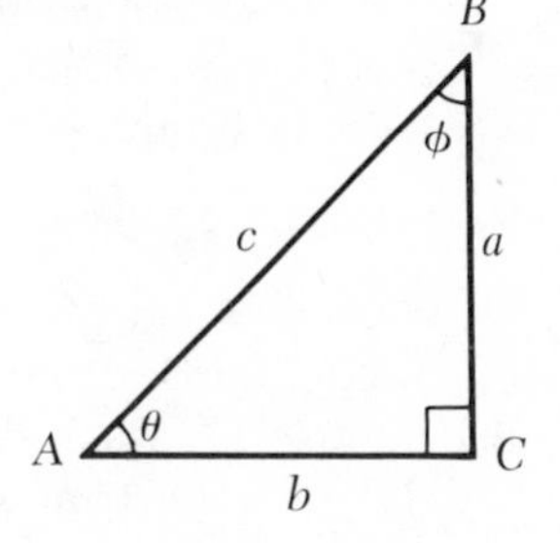

FIGURE 3.9 Right triangle with $\angle ACB = 90°$.

Similarly, the ratio b/c is defined to be the *cosine* of θ; that is,

$$\cos\theta = \frac{\text{side adjacent angle } \theta}{\text{hypotenuse}} = \frac{b}{c} \tag{3.7}$$

Using the Pythagorean theorem, $c^2 = a^2 + b^2$, we can express $\sin\theta$ and $\cos\theta$ exclusively in terms of a and b:

$$\sin\theta = \frac{a}{\sqrt{a^2+b^2}}; \quad \cos\theta = \frac{b}{\sqrt{a^2+b^2}} \tag{3.8}$$

Other trigonometric functions can also be defined, but the sine and the cosine are all that we will need here.

Example 3.5.1

Consider a right triangle with $a = 4$ cm and $b = 3$ cm. In this case,

$$c = \sqrt{4^2 + 3^2}\text{ cm} = \sqrt{25}\text{ cm} = 5\text{ cm}$$

and the triangle has the form shown in the accompanying figure. Therefore, from the defining equations for $\sin\theta$ and $\cos\theta$, we find

$$\sin\theta = \frac{4\text{ cm}}{5\text{ cm}} = 0.80; \quad \cos\theta = \frac{3\text{ cm}}{5\text{ cm}} = 0.60$$

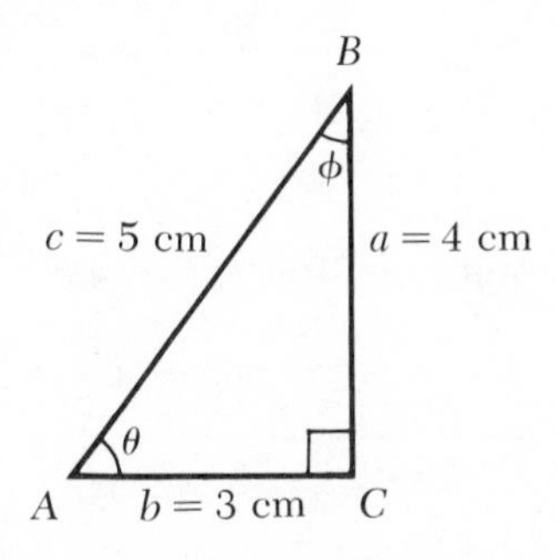

Note that the units of length cancel in the preceding expressions so that the resulting numbers for $\sin\theta$ and $\cos\theta$ are *dimensionless.*

Consider next the special (and important) triangles shown in Figures 3.10 and 3.11. As indicated in Figure 3.10, the lengths of the sides of the 45°–45°–90° triangle stand in the ratio

$$1:1:\sqrt{2}$$

For the 30°–60°–90° triangle, the lengths of the sides stand in the ratio

$$1:\sqrt{3}:2$$

as shown in Figure 3.11. Thus, it is apparent from the defining equations for the sine and cosine functions that

$$\sin 45° = \frac{1}{\sqrt{2}} = 0.71; \quad \cos 45° = \frac{1}{\sqrt{2}} = 0.71$$

and

$$\sin 30° = \cos 60° = \frac{1}{2} = 0.50$$

$$\cos 30° = \sin 60° = \frac{\sqrt{3}}{2} = 0.87$$

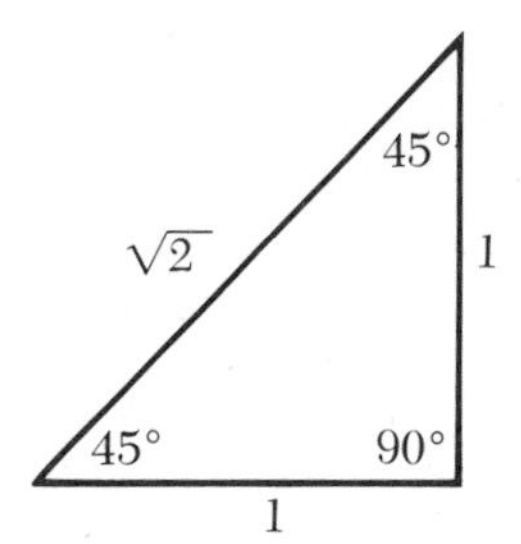

FIGURE 3.10 The 45°–45°–90° triangle.

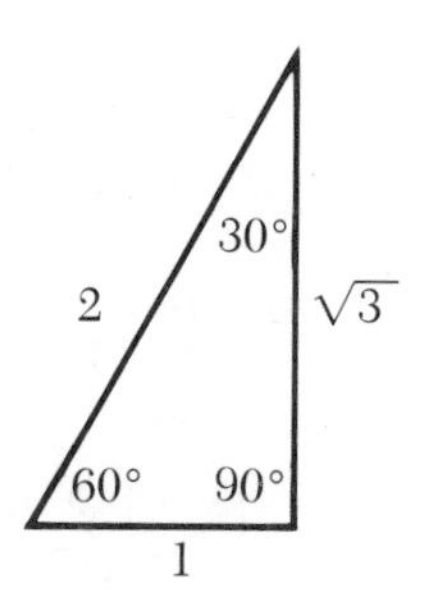

FIGURE 3.11 The 30°–60°–90° triangle.

TABLE 3.3 VALUES OF SOME TRIGONOMETRIC FUNCTIONS

	$\sin\theta$	$\cos\theta$
$\theta = 0°$	0	1
$\theta = 30°$	0.50	$\frac{\sqrt{3}}{2} = 0.87$
$\theta = 45°$	$\frac{1}{\sqrt{2}} = 0.71$	$\frac{1}{\sqrt{2}} = 0.71$
$\theta = 60°$	$\frac{\sqrt{3}}{2} = 0.87$	0.50
$\theta = 90°$	1	0

Table 3.3 summarizes the values of the trigonometric functions for all the angles involved in these simple triangles, plus the cases $\theta = 0°$ and $\theta = 90°$. Of course, Table 3.3 is by no means complete. For example, it is of no assistance in calculating cos 21°. Given in Table I (page 99) is a more extensive table of values for $\sin\theta$ and $\cos\theta$ for angles ranging from $\theta = 0°$ to $\theta = 90°$ in steps of 1°. In order to determine sin 10°, for example, we read down the left-hand column on page 99 to $\theta = 10°$, and then read across to the sine column to find

$$\sin 10° = 0.17$$

In a similar manner it is easy to verify that

$$\sin 83° = 0.99; \cos 21° = 0.93$$

and so on. (Look up these values in Table I to make certain that you understand the use of the table.)

For problems in which the value of $\sin\theta$ or $\cos\theta$ is specified, the trigonometric tables can also be used (in the reverse fashion to that described above) to evaluate the angle θ. For example, if $\sin\theta = 0.71$, then, from Table 3.3, $\theta = 45°$. Similarly, it follows from Table I that

$$\sin\theta = 0.44 \text{ corresponds to } \theta = 26°$$

$$\cos\theta = 0.67 \text{ corresponds to } \theta = 48°$$

(Again, look up these values in the table.)

Example 3.5.2

Consider the right triangle shown in the figure below. This is a particularly simple triangle because the sides stand in the ratio 3:4:5. (Notice that $5^2 = 4^2 + 3^2$). Evidently,

$$\cos \theta = \frac{3}{5} = 0.60$$

From Table I, page 99, we conclude that $\theta = 53°$ (approximately). Moreover, $\phi = 90° - \theta = 37°$.

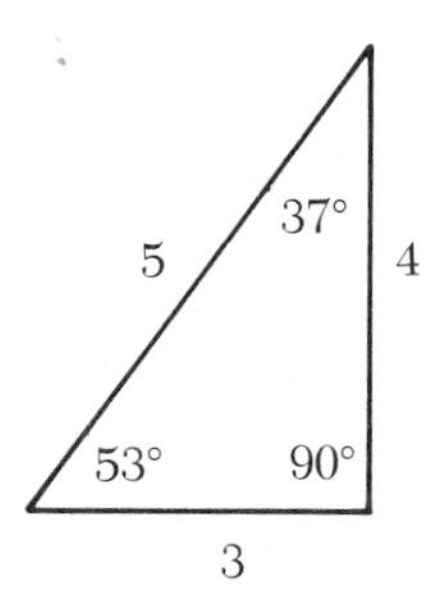

Example 3.5.3

A surveyor wishes to determine the distance between two points A and B, but he cannot make a direct measurement because a river intervenes. How can he obtain a precise value for the distance?

The surveyor first selects a point C and adjusts the location of C until a sighting with his transit (located at A) shows that the lines AB and AC are at right angles: that is, $\angle CAB = 90°$. He then measures the distance from C to A and finds

$$CA = 264 \text{ ft}$$

Next, he positions his transit at C and sights toward A and then toward B, measuring the angle between the lines CA and CB. He finds $\angle ACB = 62°$.

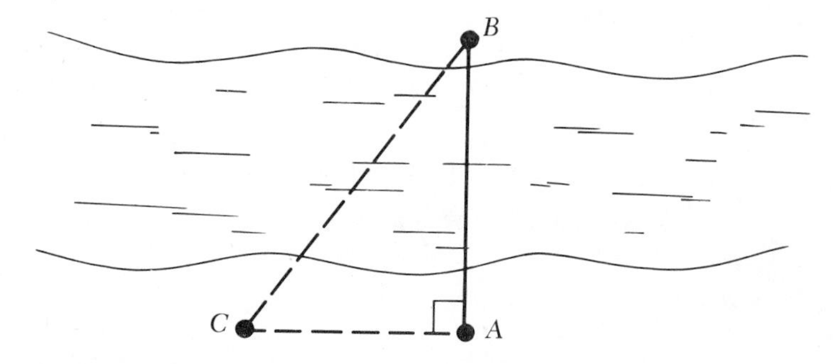

According to the definition of the cosine function, we can write

$$\cos \angle ACB = \frac{CA}{CB}$$

Solving for CB,

$$\begin{aligned} CB &= \frac{CA}{\cos \angle ACB} \\ &= \frac{264 \text{ ft}}{\cos 62^\circ} \\ &= \frac{264 \text{ ft}}{0.47} \\ &= 562 \text{ ft} \end{aligned}$$

Next, we write the expression for the sine of $\angle ACB$:

$$\sin \angle ACB = \frac{AB}{CB}$$

Solving for AB,

$$\begin{aligned} AB &= CB \cdot \sin \angle ACB \\ &= (562 \text{ ft}) \cdot (\sin 62^\circ) \\ &= (562 \text{ ft}) \cdot (0.88) \\ &= 495 \text{ ft} \end{aligned}$$

Notice that the actual distance from A to C does not matter. That is, as long as $\angle CAB = 90^\circ$ and the points A and B are fixed, the location of point C is irrelevant. The surveyor needs only to measure the distance AC and the angle ACB; any combination will then produce the same result for AB.

Notice also that we solved the problem in two steps: first, we found the length of the hypotenuse CB and then used this result to obtain the desired distance AB. We could have solved the problem in a single step if we had introduced and used the *tangent* function. This is a slightly simpler procedure but it is not necessary.

EXERCISES

Consider the right triangle in Figure 3.9. If $a = b = 1$ ft, then

1. $c =$ ____ (Ans. 115)
2. $\sin \theta =$ ____ (Ans. 116)
3. $\cos \theta =$ ____ (Ans. 117)

Consider the right triangle in Figure 3.9. If $b = 6$ m and $c = 10$ m, then

4. $a =$ ____ (Ans. 118)

5. $\sin \theta =$ ____ (Ans. 119)

6. $\cos \theta =$ ____ (Ans. 120)

(Hint: Since b and c are known, a can be determined by writing the Pythagorean theorem in the form $a^2 = c^2 - b^2$).

7. $\sin 1° =$ ____ (Ans. 121)

8. $\cos 72° =$ ____ (Ans. 122)

9. $\sin$ ____ $= 0.77$ (Ans. 123)

10. $\cos$ ____ $= 0.88$ (Ans. 124)

CHAPTER FOUR

GRAPHS

4.1 TYPES OF GRAPHS

In many situations in the physical sciences the easiest way to convey information clearly is by use of a graph. Many types of graphical presentations are possible. One of the simplest is the "pie" graph shown in Figure 4.1. The entire "pie" represents 100 per cent of some quantity, and the area of each "slice" indicates the fraction devoted to or consumed by a particular element of the whole. A pie graph gives *relative* information. For example, we can see in Figure 4.1 that the expenditures for "procurement" and for "personnel" were approximately the same

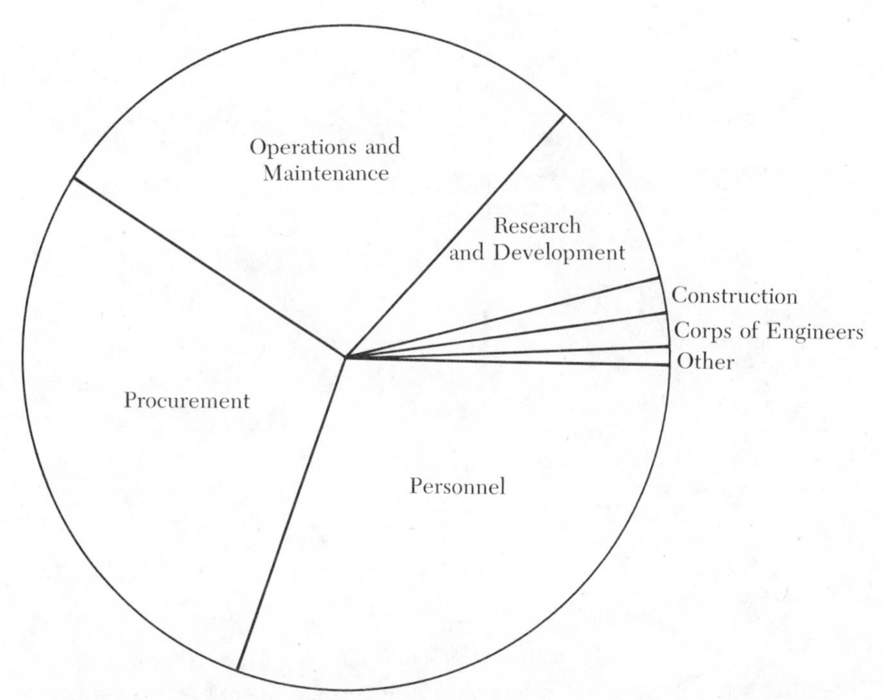

FIGURE 4.1 A "pie" graph showing the fraction of the U.S. Department of Defense budget spent for various purposes in Fiscal Year 1970 (July 1, 1969, to June 30, 1970). The total DOD expenditures were approximately 90 billion dollars ($\$90 \times 10^9$).

for Fiscal Year 1970 (FY70), but we cannot obtain the dollar figures (the *quantitative* information) unless the fractional size of each slice is given or is measured and unless the total value of the pie is known. In the case of Figure 4.1, the total value of the pie is given in the caption, and the relative size of each segment can be measured with a protractor. Therefore, sufficient information is available to permit an approximate computation of the allocation of funds by the Department of Defense in FY70.

Figure 4.2 shows a pie graph in which the relative sizes of the segments are indicated on the diagram. This is the usual practice when one (or more) of the slices is so small that it cannot be indicated with its actual size. In Figure 4.2, for example, if the atmospheric contribution were drawn with its actual size, it would be invisible.

Figure 4.3 shows a different kind of graph. Here, the vertical scale provides quantitative information—it represents the area of a lake in square miles. The horizontal direction, however, has no significance; each oblong box refers to one lake, in decreasing order of size. From this graph we can see that the Caspian Sea (which is actually a lake) is by far the largest inland body of water in the world with an area of approximately 144 000 sq mi. Lake Michigan, with an area of approximately 22 000 sq mi, is the largest lake that lies entirely within the United States.

The next degree of complexity in a graph is to allow both the vertical and horizontal directions to carry quantitative information. Figure 4.4 shows the growth of the world population from about 1600 to the present

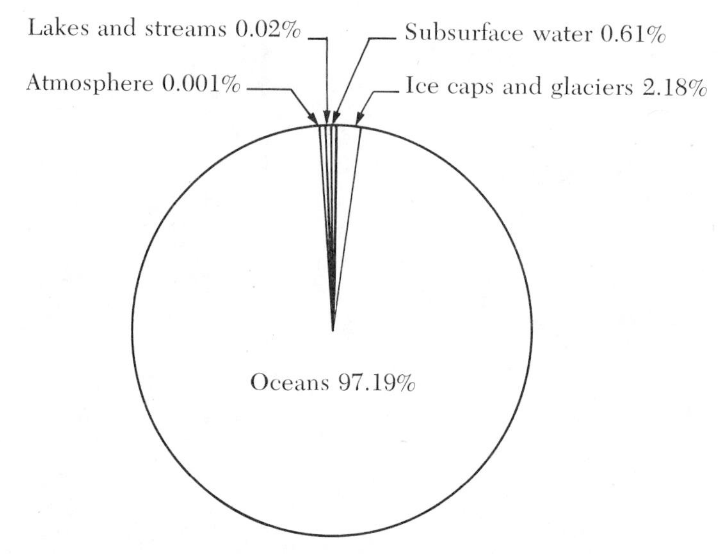

FIGURE 4.2 A "pie" graph showing the distribution of water over the Earth. (The sizes of the "pie slices" for the smallest segments have been exaggerated.)

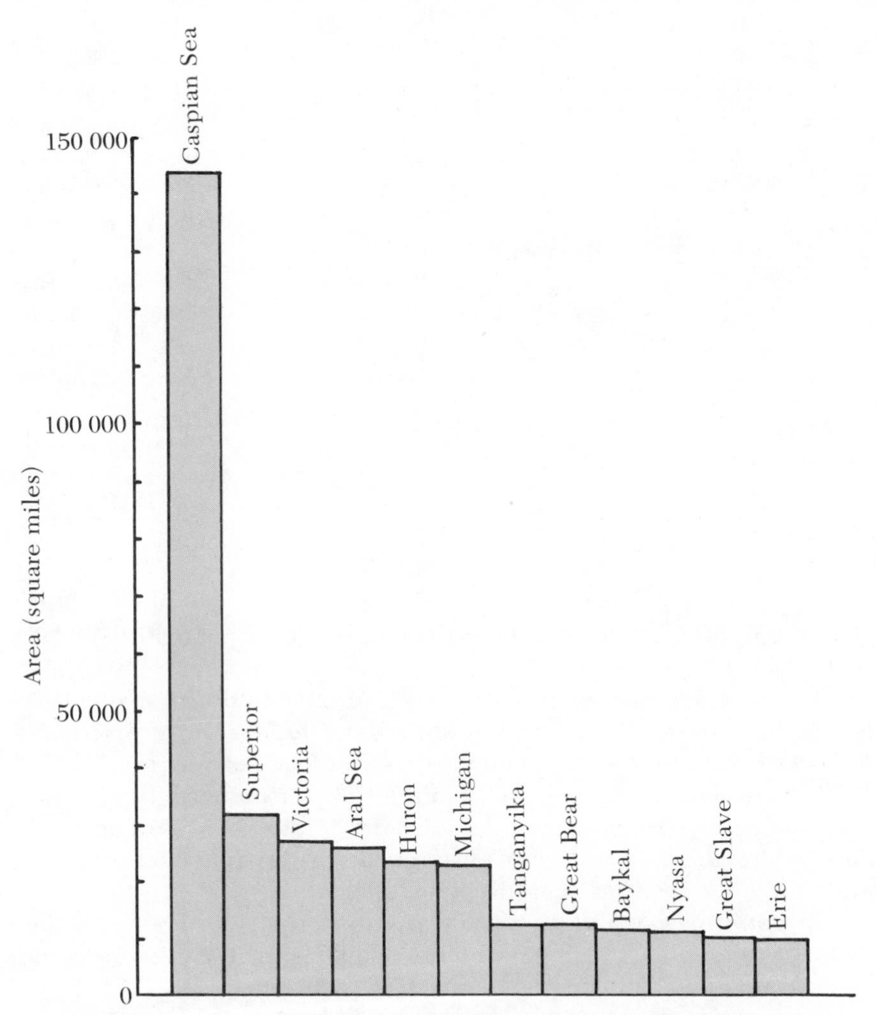

FIGURE 4.3 Comparison of the areas of the dozen largest lakes in the world.

time. The vertical scale indicates the population in billions of people and the horizontal scale gives the time in years A.D. A graph such as this is imprecise for two reasons. First, the quantity of interest (the world population at a particular time) is never known with high accuracy. Second, even if the population figure were known exactly, it could not be read from a graph that necessarily compresses a huge set of numbers into a small range. To read an *approximate* value of the world population at a particular time (for example, the year 1900), we construct a vertical line from the position 1900 on the horizontal scale, as shown in Figure 4.4. From the point at which this vertical line meets the curve, we draw a horizontal line to the vertical scale at the left. This line meets the scale between the positions marked 1 and 2 billion people. In this case we estimate that the horizontal line corresponds to a scale value of 1.6.

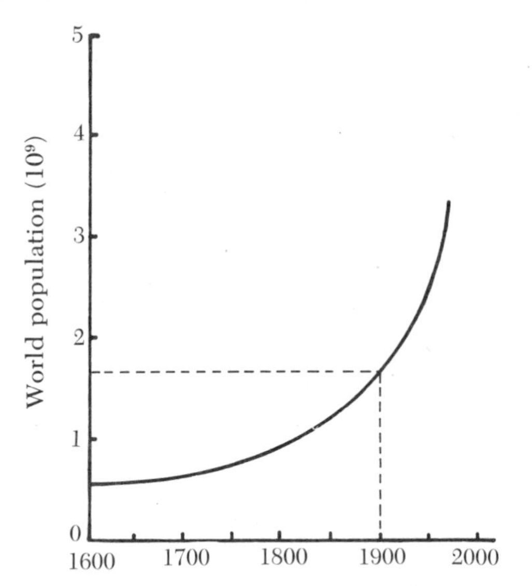

FIGURE 4.4 The world population from 1600 to the present.

We therefore conclude that the approximate world population in 1900 was 1.6 billion.

Figure 4.5 is another example of quantitative information displayed in graphical form. This diagram shows the number of eruptive disturbances (*sunspots*) counted on the Sun's surface per year from 1860 to 1970. Each dot corresponds to a definite integer number (there are no fractional sunspots), and each dot is connected to its immediate neighbors by straight lines. The only purpose of these straight lines is to assist in guiding the eye from point to point.

It would be possible to plot these sunspot data with an expanded vertical scale so that the precise number could be read for each year. But in this situation the important aspect of solar activity to show in the graph is not the exact number of sunspots. Instead, it is more interesting to concentrate on the way in which the number of sunspots changes from year to year. Here we see the regular series of increases and decreases in sunspot numbers. The maximum numbers of sunspots in the "active Sun" years are not the same from peak to peak, but the time interval from maximum to maximum averages about 11 years. This is called the *sunspot cycle.* The period 1974 to 1976 are "quiet Sun" years. When will the next "active Sun" period occur?

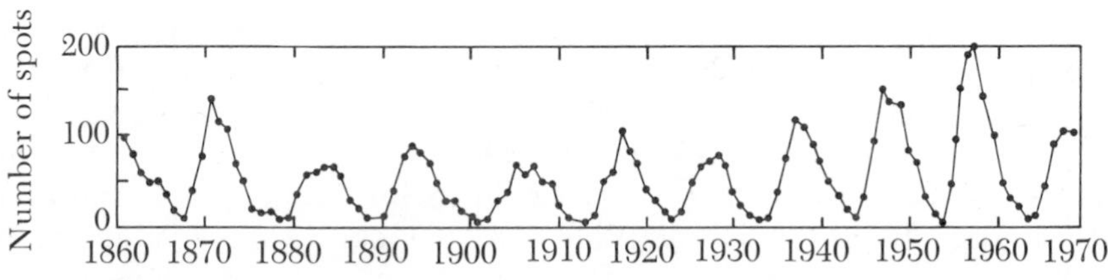

FIGURE 4.5 Variation of the number of sunspots by year from 1860 to 1970.

EXERCISES

1. Refer to Figure 4.1. The fraction of the total area of the "pie" occupied by one of the "slices" can be obtained by measuring the angle between the two straight lines that define the slice and taking the ratio of this angle to 360° (which equals the complete circle). Use a protractor to measure the angle in the "procurement" section and estimate the total expenditure for this item in FY70. (Ans. 125)

2. Use the information in Figure 4.2 and determine the volume of water contained in the ice caps and glaciers. (Ans. 126)

3. Estimate the ratio of the area of Great Slave Lake (Canada) to that of the Caspian Sea. (Use Fig. 4.3.) (Ans. 127)

4. What was the approximate world population at the time of the American Civil War? (Fig. 4.4.) (Ans. 128)

5. Using Figure 4.5, tabulate the time intervals between successive sunspot maxima for each cycle from 1870 to 1970. What do you obtain for the average value? (Ans. 129)

6. Make a bar graph showing the maximum and minimum temperatures in your locality for a period of one or two weeks. Compute the "average high" and the "average low."

4.2 RECTANGULAR COORDINATES

Various kinds of physical information are often presented in a form similar to that shown in Figures 4.4 and 4.5, namely, on a grid with values indicated on the vertical and horizontal axes. A pair of perpendicular *coordinate axes* and a *scale* for each are the essential ingredients of a *rectangular coordinate* system. It is customary to use X for the horizontal axis (or *abscissa*) and Y for the vertical axis (or *ordinate*), and to allow both positive and negative values for each coordinate. Such a coordinate system is illustrated in Figure 4.6.

The location of a point in the X-Y coordinate system is specified by stating two numbers, the value of the X coordinate and the value of the Y coordinate; for example ($x = 3$ units, $y = 4$ units). Usually, we simplify this procedure and write only the value of x and the value of y, it being understood that the order of the coordinates is first x and then y. Thus,

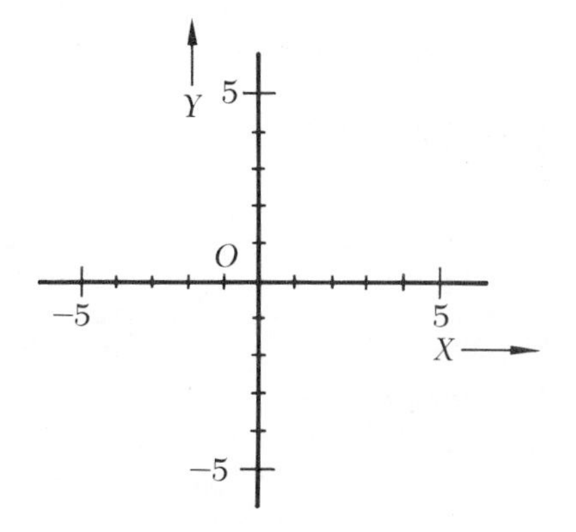

FIGURE 4.6 An *X-Y* rectangular coordinate system.

the point referred to above is written as (3,4); a general point on the plane is written as (x,y). The origin is denoted by (0,0). Figure 4.7 locates several points in this notation.

Notice that we have not specified the magnitude of a unit of measure along the X and Y axes. Each unit could represent 1 cm, or 1 mile, or even 5.8 miles; in fact, the X unit could be different from the Y unit.

In Section 3.4 we showed how to use the Pythagorean theorem to determine the length of one side of a right triangle if the lengths of the other two sides are known. Because the axes of rectangular coordinate system are at *right* angles, the Pythagorean theorem can be used to calculate the distances between pairs of points in such a system. Consider the point A with coordinates (3,4) shown in Figure 4.8. How far is

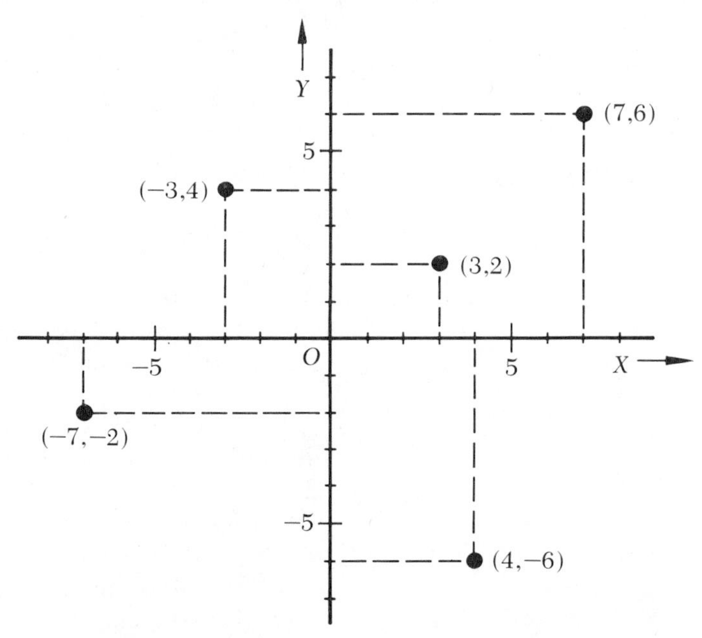

FIGURE 4.7 The location of points *(x,y)* in the *X-Y* plane.

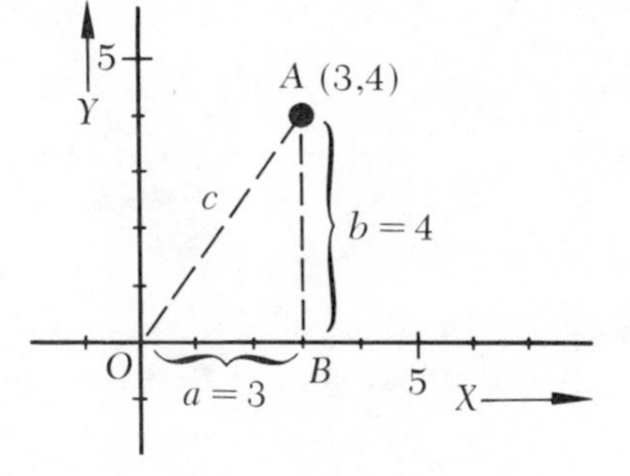

FIGURE 4.8 The distance c from O to A can be found by using the Pythagorean theorem.

A from the origin? The points A, B, and the origin O define the right triangle OBA. Therefore, the distance c from O to A is

$$c = \sqrt{a^2 + b^2} = \sqrt{(3)^2 + (4)^2} = \sqrt{9 + 16}$$
$$= \sqrt{25} = 5 \text{ units}$$

Notice that this procedure works even if the coordinates of A are *negative* values because it is the *squares* of the coordinates that enter into the calculation.

If we wish to find the distance between two points neither of which is the origin, then we follow a similar procedure. Such a case is shown in Figure 4.9, in which we require the distance from A to B. The two points A and B and the point C (which has the same x-value as B and the same y-value as A) define the right triangle ACB. Then, $c = \sqrt{a^2 + b^2}$, where

$$a = x_2 - x_1; \quad b = y_2 - y_1 \tag{4.1}$$

Therefore,

$$c = \sqrt{(x_2 - x_1)^2 + (y_2 - y_1)^2} \tag{4.2}$$

In Figure 4.9 the coordinates of the points are

$$A\text{: } (x_1,y_1) = (4,2); \quad B\text{: } (x_2,y_2) = (8,7)$$

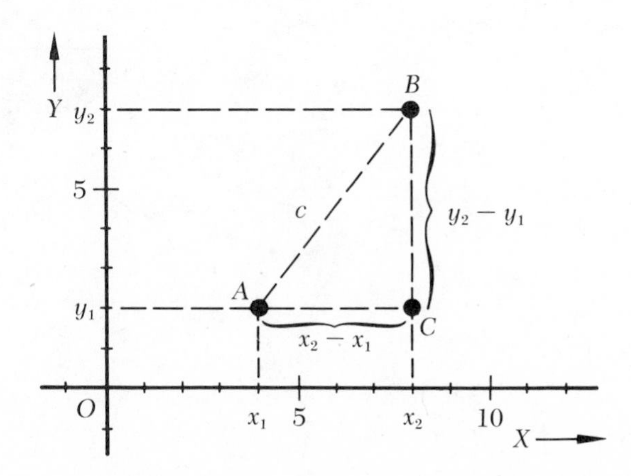

FIGURE 4.9 What is the distance from A to B?

Thus,

$$c = \sqrt{(8-4)^2 + (7-2)^2} = \sqrt{16+25}$$
$$= \sqrt{41} = 6.4 \text{ units}$$

EXERCISES

Plot the following sets of points on X-Y graphs and identify the geometrical shapes that the points outline:

1. (0,0), (2,6), (3,9), (5,15) (Ans. 130)
2. (4,−2), (−1,3), (−1,−2), (4,3) (Ans. 131)
3. What is the length of the diagonal of a square whose area is 8 cm^2? (Ans. 132)
4. What is the distance between the points (−3,7) and (−4,−1)? (Ans. 133)
5. What is the distance from the origin to the point (6,5)? (Ans. 134)

4.3 GRAPHS AND EQUATIONS

Suppose that we observe the motion of a certain object and record the position at various instants of time. We might obtain a set of data such as that shown in Table 4.1. Notice that we clearly label each column with the *symbol* for the quantity (t and x) and with the *unit* for the quantity (seconds and meters). This procedure should always be followed in tabulating data.

In order to plot the data in Table 4.1, we choose a rectangular coordinate system with time measured along the horizontal axis and displacement measured along the vertical axis. Thus, we have a t-x graph. (Not every graph is an x-y graph!) Notice in Figure 4.10 that convenient scales have been included along the axes and that the units (*meters* for

TABLE 4.1 MOTION OF AN OBJECT

Time t (seconds)	Distance Moved x (meters)
0	0
1	15
2	30
3	45
4	60
5	75
6	90
7	105

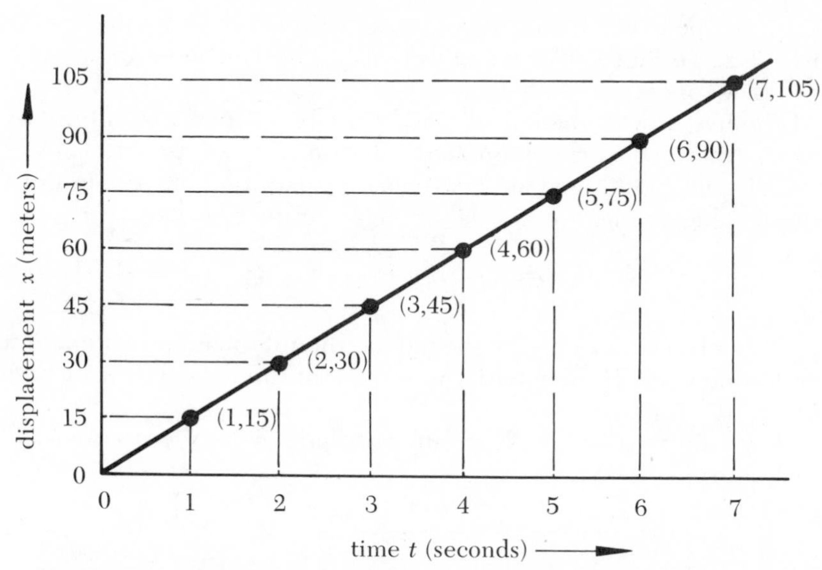

FIGURE 4.10 Graph of the data in Table 4.1 representing the motion of an object.

displacement and *seconds* for time) have been clearly indicated. This procedure should always be followed when plotting data.

To plot the data points which are tabulated in Table 4.1, we proceed as follows:

At $t = 0$ the value of x is $x = 0$. The coordinates of this point are (0,0), which is simply the *origin* in Figure 4.10.

According to Table 4.1, at $t = 1$ s the value of x is $x = 15$ m. The coordinates of this point are (1, 15). As indicated by the first pair of dotted lines in Figure 4.10, the point (1, 15) is located 1 s along the horizontal axis, and 15 m along the vertical axis.

From the table at $t = 2$ s the value of x is $x = 30$ m. The coordinates of this point are (2, 30). As indicated by the second pair of dotted lines in Figure 4.10, the point (2, 30) is located 2 s along the horizontal axis, and 30 m along the vertical axis.

In a similar manner, the data points (3, 45), (4, 60), (5, 75), (6, 90), and (7, 105) are plotted in Figure 4.10.

In Figure 4.10, we have connected the data points by a solid line. Because a *single straight line* passes through *all* data points, we say that x and t are connected by a *linear relationship*. In this example, the displacement x increases *linearly* with time t at a *constant* rate of 15 meters per second. That is, the object moves with a constant speed, $v = 15$ m/s.

We can express the conclusion we have reached regarding the motion of the object in terms of an *equation*. Although a graph clearly presents a visual picture of the situation, it is often more convenient to summarize the data in a compact mathematical statement by means of an equation. All of the data in Table 4.1 and Figure 4.10 can be represented by a simple equation,

$$x = vt \tag{4.3}$$

where the speed v is 15 m/s. When we substitute $t = 3$ s, we find $x = 45$ m; for $t = 5$ s, we find $x = 75$ m; and so forth. These values are the same as those that appear in Table 4.1.

The use of an equation instead of a table of numbers allows us to find the value of x for untabulated values of t. If we need to know x at $t = 3.17$ s, we could read an approximate value from the graph. But the precise value can be found easily by using Equation 4.3:

$$x = (15 \text{ m/s}) \cdot (3.17 \text{ s}) = 47.55 \text{ m}$$

(It is not always possible to reduce the information in a graph to a simple equation. How would you represent the data shown in Figure 4.5?)

Any straight line graph in an x-y plot can be represented by an equation of the form

$$y = ax \tag{4.4}$$

if the line passes the origin, as in Figure 4.10. The constant a is called the *slope* of the straight line and determines how rapidly y increases with x. Figure 4.11 shows three straight lines of the form $y = ax$ with different values of a. Notice that the larger the value of a, the steeper is the straight line.

If the straight line in a graph does not pass through the origin, then we must add a constant to Equation 4.4 and write

$$y = ax + b \tag{4.5}$$

This is the most general equation for a straight line.

Figure 4.12 shows three straight lines. All of the lines have the same slope, but only line ② passes through the origin. Looking at line ②, we

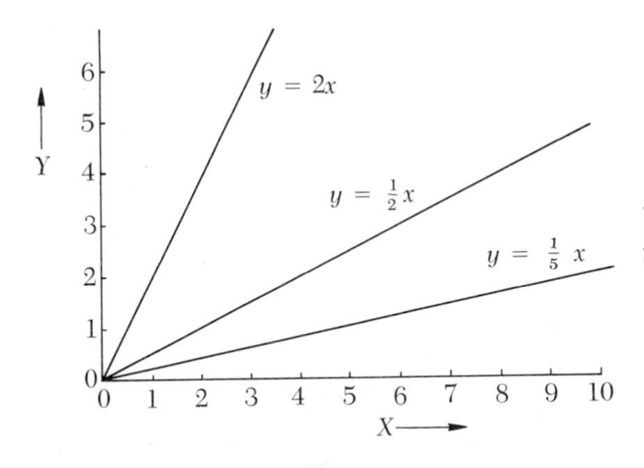

FIGURE 4.11 Graphs of three straight lines of the form $y = ax$ with different values of the slope a.

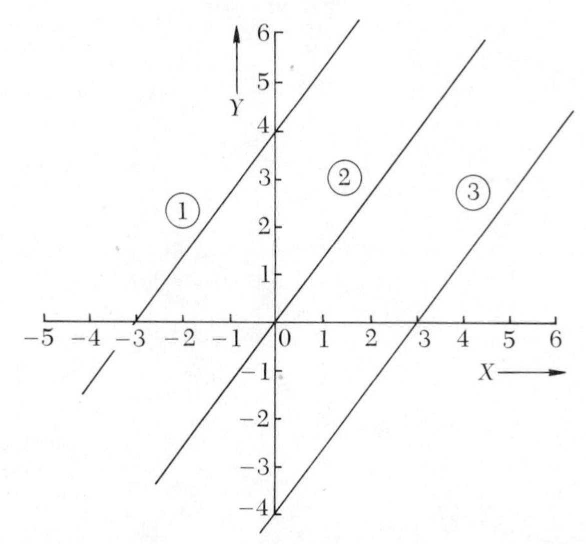

FIGURE 4.12 Three straight lines with the same slopes but with different values of *b* in the equation $y = ax + b$.

see that y increases by 4 units as x increases by 3 units. Therefore, the slope of the line is 4/3, and we can write

$$\text{Line ②: } y = \frac{4}{3}x$$

Line ① has the same value of a as line ②, but when $x = 0$, $y = 4$ in this case. Therefore, the constant b in Equation 4.5 is equal to 4 and we have

$$\text{Line ①: } y = \frac{4}{3}x + 4$$

Similarly, line ③ intersects the Y-axis at $y = -4$, so the equation for this line is

$$\text{Line ③: } y = \frac{4}{3}x - 4$$

Equation 4.5 is a *linear* equation: y depends on the *first power* of x and the graph of the equation is a *straight line*. If we write an equation in which y depends on the *second power* (or the *square*) of x, we call this a *quadratic* equation (see Section 2.3). The simplest quadratic equation has the form

$$y = ax^2 \tag{4.6}$$

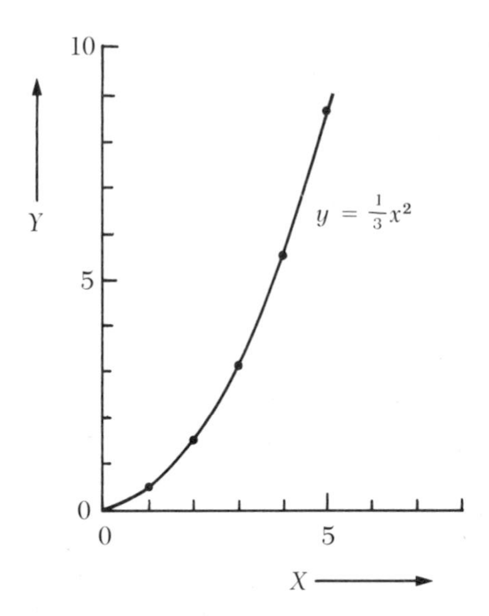

FIGURE 4.13. Graph of the quadratic equation, $y = \frac{1}{3}x^2$. The curve is called a *parabola*.

Figure 4.13 shows a graph of the equation $y = \frac{1}{3}x^2$. A curving line of this shape is called a *parabola*. We encounter curves of this type in the discussion of uniformly accelerated motion. For example, the distance x through which an object will fall in a time t, starting from rest at $x = 0$, is given by $x = \frac{1}{2}gt^2$, where g is the acceleration due to gravity. This equation has the same form as Equation 4.6 and the t-x graph of the motion is similar to the curve in Fig. 4-13.

Exercises

1. Make a t-x graph of the motion of an object that travels with a constant speed of 8 m/s and passes through the position $x = 0$ at $t = 0$. What is the equation that describes the motion? (Ans. 135)

2. Rework Exercise 1 for the case in which the object passes through the position $x = 6$ m at $t = 0$. (Ans. 136)

3. Rework Exercise 1 for the case in which the object passes through the position $x = 6$ m at $t = 2$ s. (Ans. 137)

4. Make a graph showing the two equations, $y = 2x + 3$ and $y = 3x + 2$. At what point do the two lines intersect? (Ans. 138)

5. When an object falls toward the Earth from a position near the Earth, the acceleration is $g = 32$ ft/s^2 and the equation describing the

motion is $x = \frac{1}{2}gt^2 = 16\ t^2$, if t is measured in seconds and x in feet. Make a graph of the distance of fall versus time for the interval $t = 0$ to $t = 4$ s. Use the graph to estimate the time for a fall of 100 ft. (Ans. 139)

4.4 SOME SPECIAL GRAPHS

One of the familiar shapes with which we are all acquainted is that of a *wave*. We recognize the regular rises and falls associated with the motion of waves. We see this pattern of motion in water waves and in waves that move along a stretched string or spring. Actually, the shape of an ideal wave is the same as that of the *sine* function, which was described in Section 3.5. If we look at Table I, page 99, we see that the value of sin θ starts at zero for $\theta = 0°$ and rises smoothly to 1 for $\theta = 90°$. If we were to continue the tabulation of sin θ values, we would find that for the range of angles from $\theta = 90°$ to $\theta = 270°$, sin θ decreases from $+1$, becoming zero at $\theta = 180°$ and -1 at $\theta = 270°$. Beyond $\theta = 270°$, sin θ again increases and is equal to zero at $\theta = 360°$ (which is the same angle as $\theta = 0°$).

Figure 4.14 shows a graph of sin θ versus θ for angles from $-720°$ to $+720°$. Notice that the sine function repeats itself every 360°: the value for $\theta = 90°$ is the same as that for $\theta = -270°$ or $\theta = 450°$, and so forth. That is, 360° corresponds to *one cycle* of the sine function; in the case of wave motion, this range corresponds to one cycle of the wave.

One of the problems that we sometimes encounter in the physical sciences is how to represent some quantity that continues to increase (or decrease) with time. For example, Figure 4.15 shows the way in which the consumption of electrical energy has grown in the U.S. during the last few decades. This steeply rising curve expresses the fact that in our expanding economy we have demanded ever-increasing amounts of electrical energy to support our industrial growth and our improving standard of living. In order to gauge our fuel requirements in the years

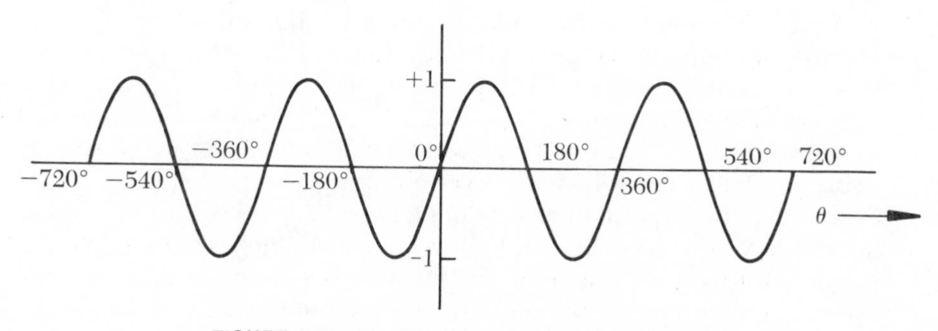

FIGURE 4.14 Sin θ for θ between $-720°$ and $+720°$.

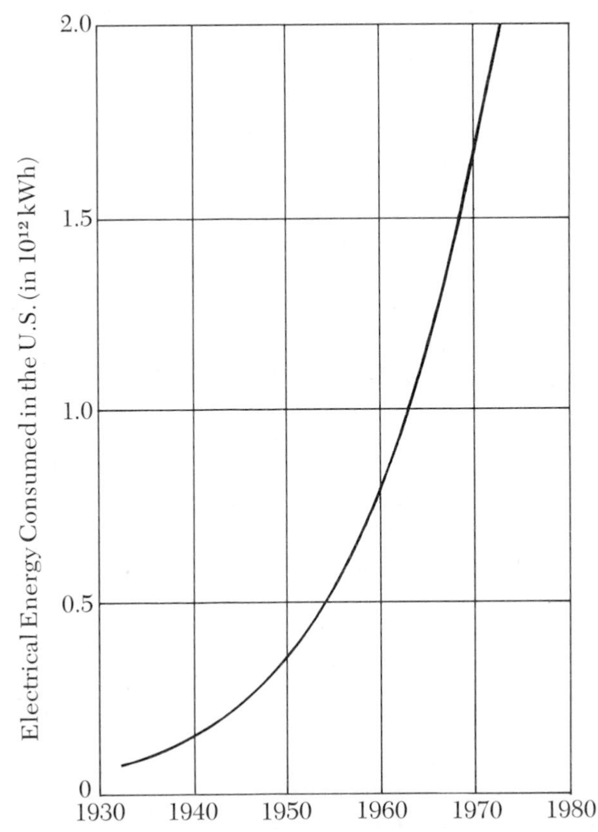

FIGURE 4.15 The increase in the consumption of electrical energy in the U.S. during the period from 1933 to 1972.

ahead, we need to project this curve into the future. But how can we extend such a steeply rising curve to future times? The curve in Figure 4.15 is *not* a parabola and so we cannot use Equation 4.6 to compute the energy consumption at various times. (Try to estimate the electrical energy demands in the year 2000 by projecting the curve in Figure 4.15.) One way to estimate a future value is to plot the information in Figure 4.15 on a different type of graph. Figure 4.16 shows the U.S. electrical energy consumption data again. Notice, however, that the vertical scale is not the same as in Figure 4.15 or in our other graphs. Here, each major division of the scale corresponds to an increase by a factor of 10. The importance of this kind of graph is that the particular type of steeply rising curve in Figure 4.15 is converted into a *straight line*. Now, if we assume that the pattern of growth of electrical energy consumption during the next few decades will be the same as in the recent past (this is not necessarily a valid assumption but it is the simplest guess we can make), then the graph can be easily extended to the desired date in the future. The dotted portion of the line in Figure 4.16 shows such an *extrapolation*.

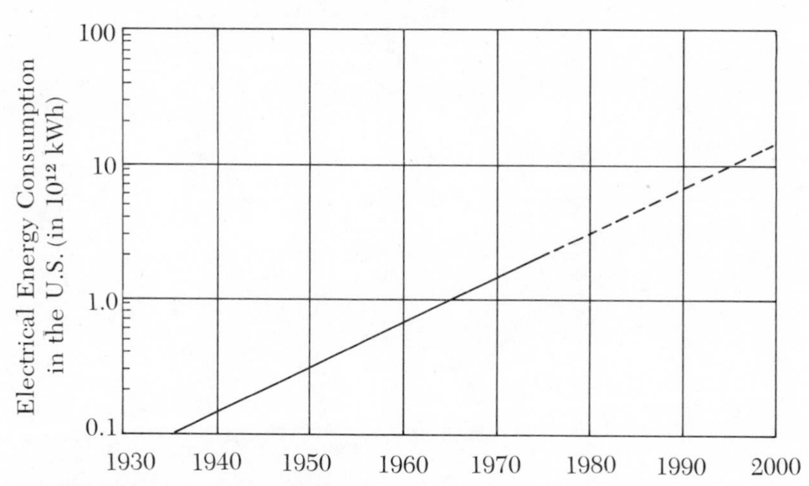

FIGURE 4.16 Logarithmic graph of the consumption of electrical energy in the U.S. This graph contains the same information as Figure 4.15, but the use of the logarithmic scale converts the curve into a straight line.

A graph in which equal intervals of the scale represent increases by a factor of 10 (or some other number) are called *logarithmic* graphs. If a quantity appears as a straight line in a logarithmic graph, as in Figure 4.16, the quantity is said to have an *exponential* increase (or decrease). In Section 6.8 we will see another application of a logarithmic graph when we discuss the exponential decay of a radioactive substance.

EXERCISES

1. Use the information in Table I, page 99, and plot the values of cos θ versus θ for $\theta = 0°$ to $\theta = 90°$ using intervals of 5°. Do you see the similarity with the graph of sin θ (Fig. 4.14)? What is the difference? Extend the graph by sketching the curve for the range $\theta = -360°$ to $\theta = +360°$. What is the value of cos θ for $\theta = -180°$? (Ans. 140)

2. The *doubling time* of an exponentially increasing quantity is the time interval required for the quantity to double its value. In every such time interval, no matter what initial time is chosen, the value of the quantity at the end of the interval will always be twice the value at the beginning of the interval. Test this statement using Figure 4.16 and find the doubling time for the consumption of electrical energy in the U.S. (Ans. 141)

CHAPTER FIVE

VECTORS

5.1 EXAMPLES OF SCALARS AND VECTORS

Many of the quantities that we use in the physical sciences can be completely described in terms of a *magnitude* only. For example, if we say that the temperature is 20°C, this is the complete specification of the temperature. Other physical quantities that require only a magnitude are mass, pressure, volume, density, time, and so forth. These quantities are collectively known as *scalars*. A scalar is completely specified by a *magnitude* alone (together with the appropriate units).

Some physical quantities have *direction* as well as *magnitude*. For example, to say that an automobile is moving at 40 mi/hr does not give a complete description of the motion. To be complete we must say that the automobile is moving with a velocity of 40 mi/hr in the direction northeast. *Velocity* is one of a class of quantities called *vectors* which require both magnitude and direction for their complete specification. Some other vector quantities are displacement, force, acceleration, and electric field.

It is customary to place an arrow (→) above the symbol representing a vector quantity; this is to emphasize that the vector has direction as well as magnitude. For example, the vector *velocity* is usually represented by $\vec{v}$ and the vector *displacement* is usually represented by $\vec{x}$ or $\vec{s}$. One also frequently finds vector quantities indicated by bold face type: $\vec{x} = \mathbf{x}$.

To dramatize the distinction between scalars and vectors, consider the illustration in Figure 5.1. A ball is thrown vertically upward. It rises

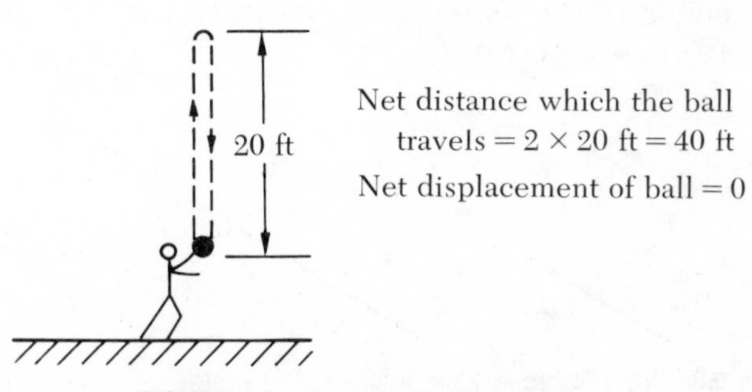

FIGURE 5.1 Zero net displacement.

to a height of 20 ft, then falls and is caught at the same location from which it was released. The net *distance* that the ball travels (a *scalar* quantity) is equal to 2×20 ft $= 40$ ft. However, since the initial and final locations of the ball are the same, the net *displacement* of the ball (a *vector* quantity) is zero.

5.2 REPRESENTATION AND SIMPLE PROPERTIES OF VECTORS

As shown in Figure 5.2, a vector $\vec{A}$ can be represented pictorially by a *directed line segment* from some origin O (the *foot* of the vector) to a point P (the *head* of the vector). The vector $\vec{A}$ is sometimes denoted by $\vec{OP}$. The direction of the straight line segment, as inferred by the direction of the arrowhead in Figure 5.2, corresponds to the *direction* of the vector $\vec{A}$. The length of the straight line segment is proportional to the *magnitude* of $\vec{A}$, in some agreed units. It is customary to denote the magnitude of $\vec{A}$ by A. That is,

$$\text{Magnitude of vector } \vec{A} = A \tag{5.1}$$

where A is a *positive* number.

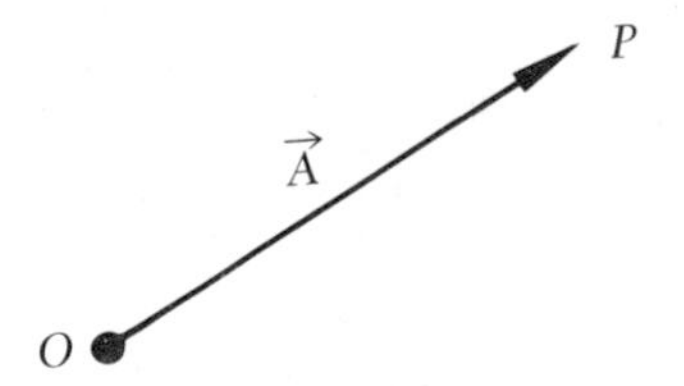

FIGURE 5.2 Pictorial representation of a vector $\vec{A}$.

If we multiply a vector by a number, we change the size or magnitude of the vector but not its direction. For example, if $\vec{A}$ stands for a vector that has a magnitude of 6 units in the direction east, then $3\vec{A}$ stands for a vector that has a magnitude of 18 units also in the direction east.

A vector that carries a *negative* sign means a vector pointing in the direction opposite to that of the corresponding vector with a positive sign. For example, the vector $-\vec{A}$ has the same magnitude as the vector $\vec{A}$, but it points in the *opposite* direction. Figure 5.3 illustrates these simple properties of vectors.

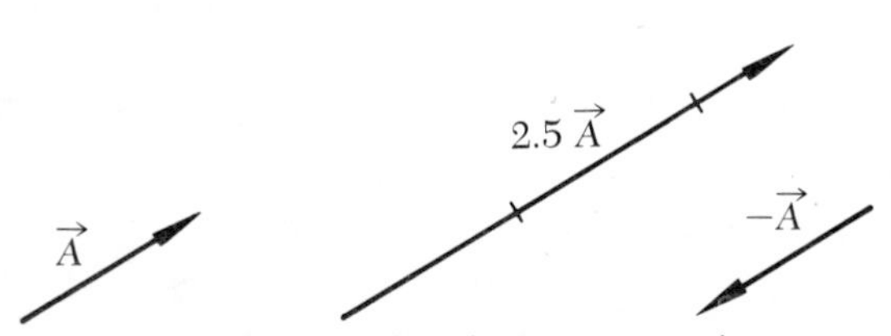

FIGURE 5.3 The vectors $\vec{A}$, $2.5\vec{A}$, and $-\vec{A}$. The vector $-\vec{A}$ is antiparallel to the other two vectors.

Example 5.2.1

As shown in the figure below, the vector $\vec{v}$ is a velocity of 50 mi/hr directed 30° north of due east. What are the magnitudes and directions of the vectors (a) $2\vec{v}$, (b) $-\vec{v}$, and (c) $-2\vec{v}$?

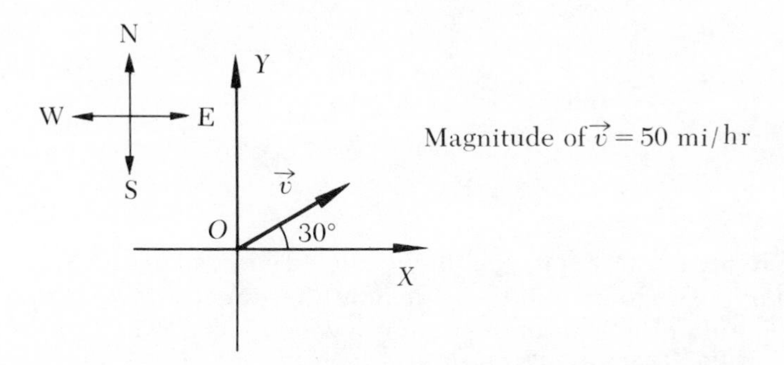

The answers to these questions are summarized pictorially in the figures to follow.

Referring to figure (a), the vector $2\vec{v}$ is in the same direction as $\vec{v}$, and has a magnitude equal to twice the magnitude of $\vec{v}$. Therefore, $2\vec{v}$ is a velocity of 2 × 50 mi/hr = 100 mi/hr directed *30° north of due east.*

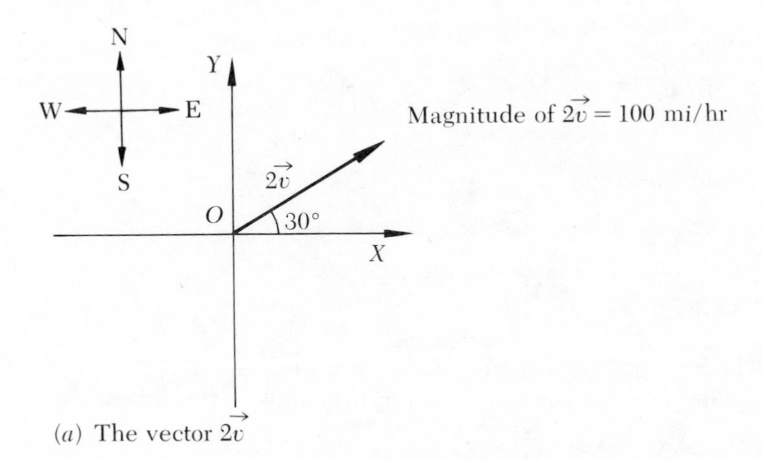

(*a*) The vector $2\vec{v}$

Referring to figure (b), the vector $-\vec{v}$ is in the direction opposite to $\vec{v}$, and has a magnitude equal to the magnitude of $\vec{v}$. Therefore, $-\vec{v}$ is a velocity of 50 mi/hr directed *30° south of due west.*

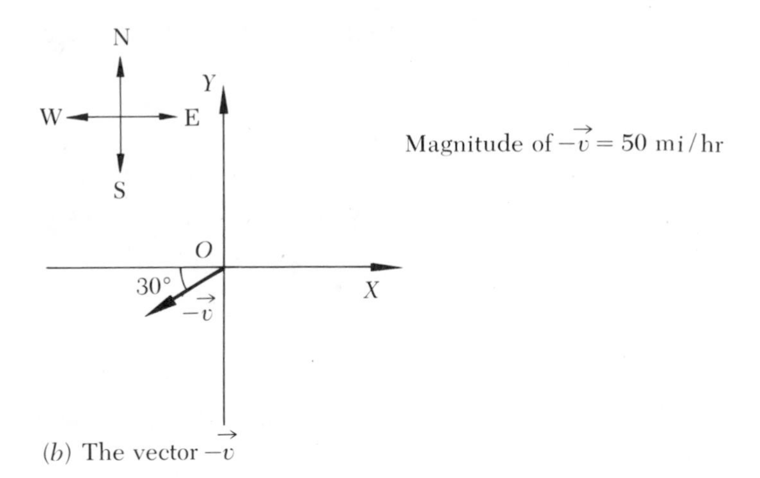

(*b*) The vector $-\vec{v}$

Referring to figure (c), the vector $-2\vec{v}$ is also in the direction opposite to $\vec{v}$, but has a magnitude equal to twice the magnitude of $\vec{v}$. Therefore, $-2\vec{v}$ is a velocity of 2×50 mi/hr= 100 mi/hr directed *30° south of due west.*

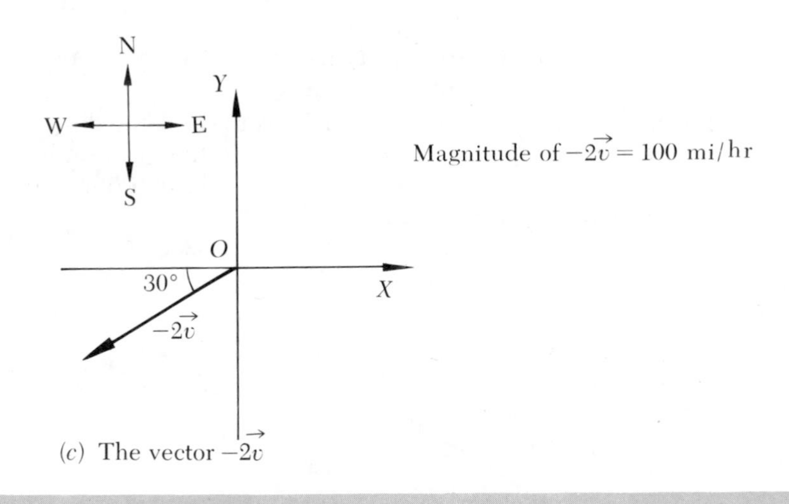

(*c*) The vector $-2\vec{v}$

Exercises

The vector $\vec{x}$ is a displacement of 25 mi directed 40° north of due west. What are the magnitudes and directions of the following vectors?

1. $6\vec{x}$ (Ans. 142)
2. $-1.5\vec{x}$ (Ans. 143)
3. $-4\vec{x}$ (Ans. 144)
4. $3\vec{x}$ (Ans. 145)

The vector $\vec{F}$ is a force of 14 newtons (N) directed vertically downward on a horizontal floor. What are the magnitudes and directions of the following vectors?

5. $5000\,\vec{F}$ (Ans. 146)

6. $-0.5\,\vec{F}$ (Ans. 147)

7. $-15\,\vec{F}$ (Ans. 148)

8. $13\,\vec{F}$ (Ans. 149)

5.3 ADDITION AND SUBTRACTION OF VECTORS

In a flat portion of rural Kansas, a motorist drives 30 miles due east from point O to point P. His *displacement* during this portion of the journey is represented by the vector $\vec{A} = \vec{OP}$ in Figure 5.4. At P the motorist turns left and then drives 40 miles due north to point P', his final destination. This displacement is represented by the vector $\vec{B} = \vec{PP'}$ in Figure 5.4. It is clear that the trip from O to P, followed by the trip from P to P', has the same end result as a trip *directly* from O to P'. As shown in the figure, the *net* displacement for the entire trip can be represented by the vector $\vec{R} = \vec{OP'}$.

The vector displacement $\vec{R}$ is called the *resultant* of the two vector displacements, $\vec{A}$ and $\vec{B}$. We can express $\vec{R}$ in equation form as

$$\vec{R} = \vec{A} + \vec{B} \tag{5.2}$$

Equation 5.2 is a *vector equation,* and states that the resultant vector displacement $\vec{R}$ is equal to the sum of the two vector displacements, $\vec{A}$ and $\vec{B}$. We note from Figure 5.4 that the vectors $\vec{A}$, $\vec{B}$, and $\vec{R}$ together form the three sides of the triangle OPP'

The magnitude and direction of the resultant vector $\vec{R}$ can be de-

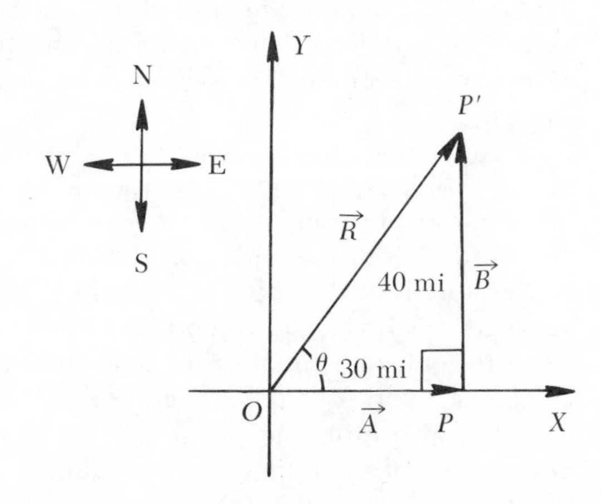

FIGURE 5.4 A journey in Kansas from O to P to P'.

termined in the following manner: Referring to Figure 5.4, and making use of the fact that OPP' is a *right* triangle with angle $\angle OPP' = 90°$, we find, from the Pythagorean theorem,

$$R = \text{magnitude of resultant vector } \vec{R} \text{ in Figure 5.4}$$
$$= \sqrt{(30)^2 + (40)^2} \text{ mi} = \sqrt{2500} \text{ mi} = 50 \text{ mi}$$

The *direction* of the resultant vector $\vec{R}$ can be determined by measuring the angle θ with a protractor; we find an angle of about 53°. Alternatively, we can make use of the definition of the cosine function (Eq. 3.7). We then write

$$\cos\theta = \frac{OP}{OP'} = \frac{30}{50} = 0.60$$

We determine θ from the table of sines and cosines on page 99, which gives $\theta = 53°$.

Therefore, we have found that the resultant vector $\vec{R}$ is equal to a displacement of 50 miles directed 53° north of due east. We emphasize that to determine completely the vector $\vec{R}$, both its *magnitude* and *direction* must be specified.

The above example illustrates a further important property of the vector addition in Equation 5.2. The vector equation $\vec{R} = \vec{A} + \vec{B}$, does not imply that the magnitude of the resultant vector $\vec{R}$ is equal to the sum of the magnitudes of the vector $\vec{A}$ and $\vec{B}$. In Figure 5.4 we see that $A + B = 30 \text{ mi} + 40 \text{ mi} = 70 \text{ mi}$, whereas $R = 50$ mi. Therefore, even though $\vec{R} = \vec{A} + \vec{B}$ is true, $R = A + B$, is *not* true. This is, in fact, a general property of vector addition, and is not restricted to the specific example illustrated in Figure 5.4.

The example discussed above also serves to demonstrate the general method for finding the resultant of any two vectors $\vec{A}$ and $\vec{B}$ which are perpendicular to one another.

We now discuss a procedure for determining the resultant $\vec{R}$ of any two vectors $\vec{A}$ and $\vec{B}$ with arbitrary orientations. The procedure is known as the *method of triangles*. Consider the vectors $\vec{A}$ and $\vec{B}$ represented pictorially in Figure 5.5 (a) and (b). The vectors $\vec{A}$ and $\vec{B}$ can signify displacements, velocities, momenta, forces, or any other type of vector.

As shown in Figure 5.5(c), we determine the resultant vector $\vec{R} = \vec{A} + \vec{B}$ by constructing a triangle in which $\vec{A}$ and $\vec{B}$ are adjacent sides and connect head to foot. The resultant vector $\vec{R}$ is then equal to the third side of the triangle. That is, the magnitude of $\vec{R}$ corresponds to the length of the third side, and the direction of $\vec{R}$ is that indicated in Figure 5.5(c). It is important to note that, while $\vec{A}$ and $\vec{B}$ connect head to foot, the direction of $\vec{R}$ is such that $\vec{A}$ and $\vec{R}$ connect foot to foot, and $\vec{B}$ and $\vec{R}$ connect head to head.

It is also important to note that the order in which $\vec{A}$ and $\vec{B}$ are connected does not affect the resultant $\vec{R}$. The two possibilities are illus-

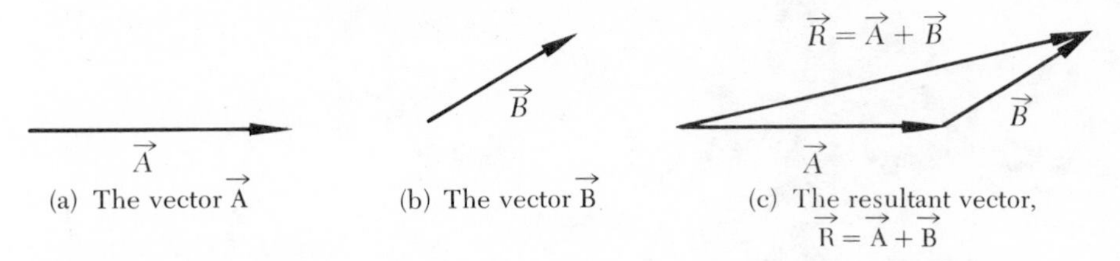

FIGURE 5.5 Constructing the triangle of vectors to determine $\vec{R} = \vec{A} + \vec{B}$.

trated in Figure 5.6. In Figure 5.6, the resultant vector $\vec{R}$ has the same magnitude and direction for both triangles. Therefore, in practice we can construct $\vec{R}$ by either method indicated in Figure 5.6. This important result is a reflection of the fact that vector addition obeys the *commutative law*; that is,

$$\vec{A} + \vec{B} = \vec{B} + \vec{A} \tag{5.3}$$

As an example that illustrates the commutative law of vector addition, we compare Figures 5.4 and 5.7. In Figure 5.4, the motorist first drives 30 miles due east, and then drives 40 miles due north to his final destination. In Figure 5.7, however, the motorist first drives 40 miles due north, and then drives 30 miles due east. As before, the 40 mile displacement northward is denoted by the vector $\vec{B}$, and the 30 mile displacement eastward is denoted by the vector $\vec{A}$.

To demonstrate that the resultant vector $\vec{R}$ is the same in both cases, we note from Figure 5.7 that $\vec{R}$ is the hypotenuse of the right triangle $OP''P'$ with $\angle OP''P' = 90°$. Therefore, from the Pythagorean theorem

$$\begin{aligned} R &= \text{magnitude of resultant vector } \vec{R} \text{ in Figure 5.7} \\ &= \sqrt{(40)^2 + (30)^2} \text{ mi} = \sqrt{2500} \text{ mi} = 50 \text{ mi} \end{aligned}$$

Next, we let θ represent the angle between $\vec{OP'}$ and the X-axis. Since $P''P'$ is parallel to the X-axis, it follows that $\angle P''P'O = \theta$. Referring to the right triangle $OP''P'$, and making use of the definition of $\cos\theta$, we find

$$\cos\theta = \frac{P''P'}{OP'} = \frac{30}{50} = 0.60$$

Again, we find $\theta = 53°$.

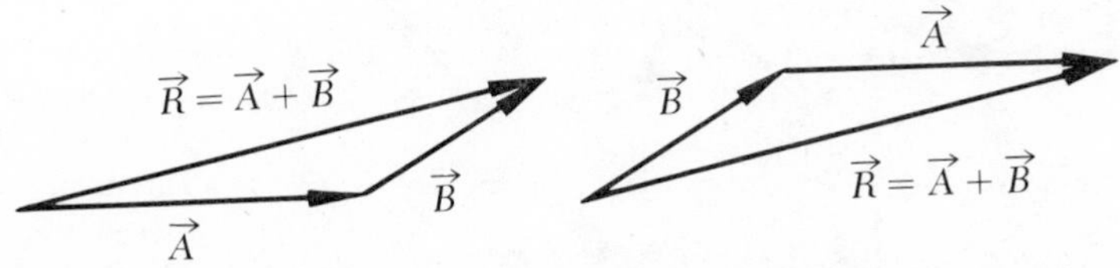

FIGURE 5.6 Two possible triangles for constructing the resultant vector, $\vec{R} = \vec{A} + \vec{B}$.

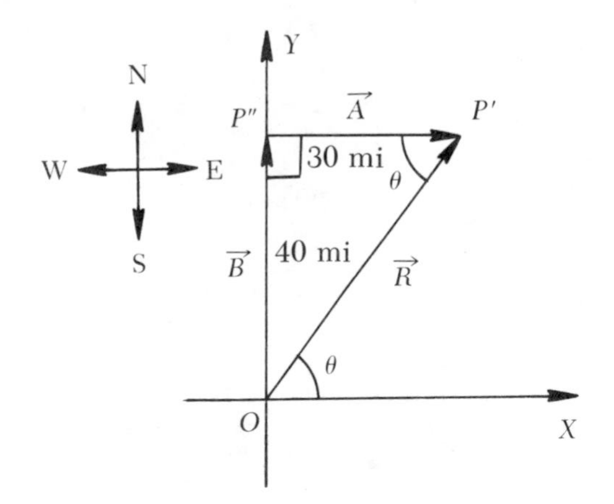

FIGURE 5.7 A journey in Kansas from O to P'' to P'.

Therefore, we have found that the resultant vector $\vec{R}$ in Figure 5.7 is equal to a displacement of 50 mi directed 53° north of due east. This is identical to the result obtained from Figure 5.4, which shows that the final destination (the point P') is the same in both cases.

The procedure for *subtracting* a vector $\vec{B}$ from a vector $\vec{A}$ is straightforward. We first reverse the direction of $\vec{B}$, that is, we form the vector $-\vec{B}$, and then *add* it to $\vec{A}$. The difference vector $\vec{D}$ is denoted by

$$\vec{D} = \vec{A} - \vec{B} \tag{5.4}$$

Since we can write $-\vec{B} = +(-\vec{B})$, Equation 5.4 can also be expressed in the form

$$\vec{D} = \vec{A} + (-\vec{B}) \tag{5.5}$$

That is, $\vec{D}$ is equal to the *sum* of the vectors $\vec{A}$ and $-\vec{B}$. As illustrated in Figure 5.8, we determine the difference vector, $\vec{D} = \vec{A} - \vec{B}$, by constructing a vector triangle in which $\vec{A}$ and $-\vec{B}$ form adjacent sides. The vector $\vec{D}$ is then equal to the third side of the triangle. That is, the magnitude of $\vec{D}$ corresponds to the length of the third side, and the direction of $\vec{D}$ is that indicated in Figure 5.8. We reiterate that the difference vector, $\vec{D} = \vec{A} - \vec{B}$, is equal to the resultant of the vectors $\vec{A}$ and $-\vec{B}$.

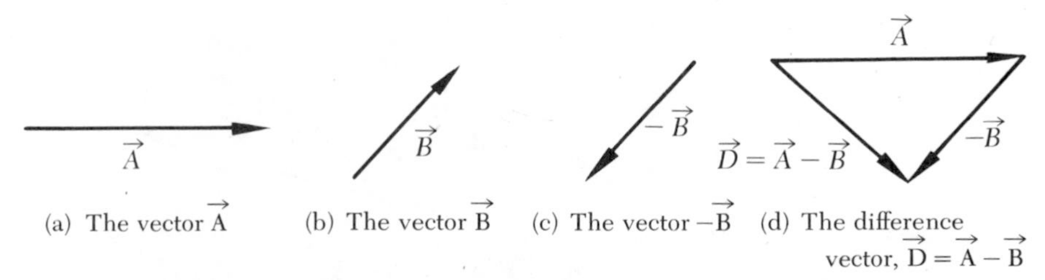

FIGURE 5.8 Constructing the triangle of vectors to determine $\vec{D} = \vec{A} - \vec{B}$.

EXERCISES

The vector $\vec{v}$ is a velocity of 60 m/s directed 45° north of due east. What are the magnitudes and directions of the resultants of the following vectors?

1. $\vec{v} + 3\vec{v}$ (Ans. 150)
2. $\vec{v} - 3\vec{v}$ (Ans. 151)
3. If $\vec{F}_1$ is a force of 2 newtons (2 N) in the positive X-direction and $\vec{F}_2$ is a force of 2 N in the positive Y-direction (see the figure), what are the magnitude and direction of the resultant force, $\vec{R}_F = \vec{F}_1 + \vec{F}_2$? (Ans. 152)
4. As shown in the figure, a pilot is attempting to fly his airplane due north with a velocity $\vec{v}$ equal to 300 mi/hr. There is a crosswind with $\vec{v}_{wind}$ equal to 100 mi/hr in the easterly direction. What are the magnitude and direction of the resultant velocity $\vec{v}_R$ of the airplane? (Ans. 153)

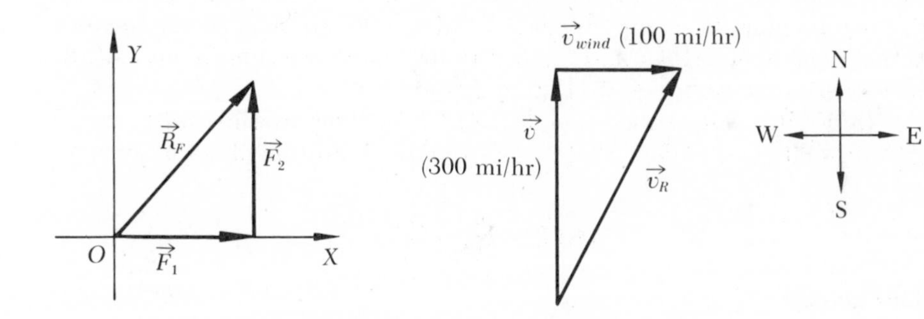

5.4 RESOLUTION OF VECTORS INTO RECTANGULAR COMPONENTS

By using the basic rules of geometry and trigonometry, any vector $\vec{A}$ can be *decomposed* into two vectors, the sum of which yields the original vector. This process is especially useful when the two vectors, which are called *component* vectors, are at right angles to one another. In this case, the two component vectors are referred to as *rectangular* component vectors. The general procedure for *resolving* a vector $\vec{A}$ into two rectangular components, $\vec{A}_x$ and $\vec{A}_y$, is illustrated in Figure 5.9.

In Figure 5.9(a), we let θ be the angle between the positive X-axis and the vector $\vec{A}$. In Figure 5.9(b), from the head of vector $\vec{A}$ (that is, the point P) we drop a perpendicular PQ to the X-axis. The angle $\angle OQP$ is

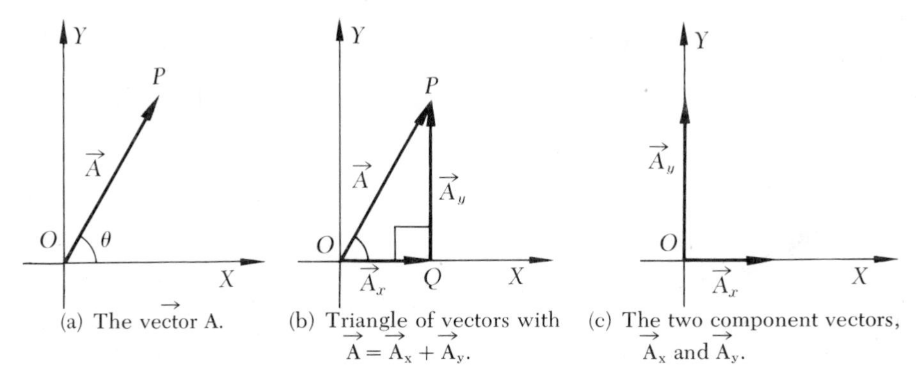

(a) The vector $\vec{A}$. (b) Triangle of vectors with $\vec{A} = \vec{A}_x + \vec{A}_y$. (c) The two component vectors, $\vec{A}_x$ and $\vec{A}_y$.

FIGURE 5.9 Resolving a vector A into two rectangular compenents, Ax and A_y: $A = A_x + A_y$.

then equal to 90°. It is evident from Figure 5.9(b) that OPQ forms a triangle of vectors consisting of $\vec{A}_x$, $\vec{A}_y$, and $\vec{A}$. Furthermore,

$$\vec{A} = \vec{A}_x + \vec{A}_y \tag{5.6}$$

That is, the vector sum of $\vec{A}_x$ and $\vec{A}_y$ is equal to the vector $\vec{A}$. The original vector $\vec{A}$ in Figure 5.9(a) can therefore be *removed* and *replaced* by the two rectangular components, $\vec{A}_x$ and $\vec{A}_y$. We emphasize that the two component vectors, $\vec{A}_x$ and $\vec{A}_y$ in Figure 5.9(c), are *totally equivalent* to the original vector $\vec{A}$ in Figure 5.9(a).

In Figure 5.9(b) the triangle OQP is a right triangle with angle $\angle OQP = 90°$, and angle $\angle POQ = \theta$. From the definition of cos θ, we find

$$\cos\theta = \frac{OQ}{OP} = \frac{A_x}{A}$$

This gives

$$A_x = A\cos\theta \tag{5.7}$$

where A is the magnitude of vector $\vec{A}$. Similarly, from the definition of sin θ, we find

$$\sin\theta = \frac{PQ}{OP} = \frac{A_y}{A}$$

This gives

$$A_y = A\sin\theta \tag{5.8}$$

Therefore, if the magnitude A of the original vector $\vec{A}$ and the angle θ which $\vec{A}$ makes with the positive X-axis are specified, the rectangular components, A_x and A_y, can be determined from the equations

$$\begin{aligned} A_x &= A\cos\theta \\ A_y &= A\sin\theta \end{aligned} \tag{5.9}$$

Referring to Figure 5.9(b) and making use of the Pythagorean theorem, we also find that

$$A = \sqrt{A_x^2 + A_y^2} \tag{5.10}$$

where A is the magnitude of the vector $\vec{A}$.

Example 5.4.1

As illustrated in the accompanying figure, the vector $\vec{F}$ represents a force of 50 newtons (N) directed at an angle of 60° with respect to the positive X-axis. What are the values of F_x and F_y?

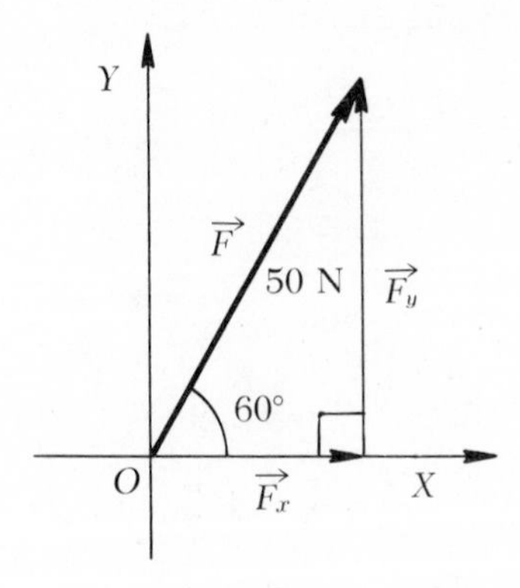

From the figure we note that

$$F = 50\ N, \text{ and } \theta = 60°$$

From Equation 5.9 and the Trigonometric Tables on page 99, we find

$$F_x = (50\ \text{N}) \cos 60° = (50\ \text{N}) \cdot (0.5) = 25\ \text{N}$$

and

$$F_y = (50\ \text{N}) \sin 60° = (50\ \text{N}) \cdot (0.87) = 43.5\ \text{N}$$

Thus, the original force vector $\vec{F}$ is equivalent to a force of 25 N in the positive X-direction, combined with a force of 43.5 N in the positive Y-direction.

Example 5.4.2

As illustrated in the accompanying figure, the vector $\vec{v}$ represented a velocity of 100 m/s directed at an angle of 45° with respect to the positive X-axis. What are the values of v_x and v_y?

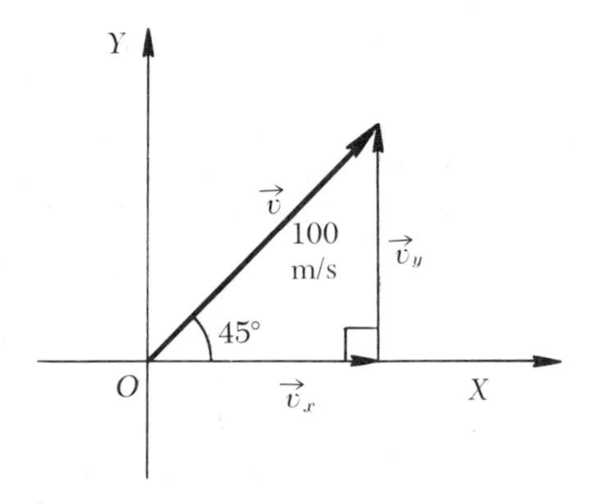

From the figure we note that

$$v = 100 \text{ m/s, and } \theta = 45°$$

From Equation 5.9 and the Trigonometric Tables on page 99, we find

$$v_x = (100 \text{ m/s}) \cos 45° = (100 \text{ m/s}) \cdot (0.71) = 71 \text{ m/s}$$

and

$$v_y = (100 \text{ m/s}) \sin 45° = (100 \text{ m/s}) \cdot (0.71) = 71 \text{ m/s}$$

Thus, the original velocity vector $\vec{v}$ is equivalent to a velocity of 71 m/s in the positive X-direction, combined with a velocity of 71 m/s in the positive Y-direction.

Example 5.4.3

As illustrated in the accompanying figure, the rectangular components of a velocity vector $\vec{v}$ are $v_x = 30$ mi/hr and $v_y =$ 40 mi/hr. What are the magnitude and direction of the vector $\vec{v}$?

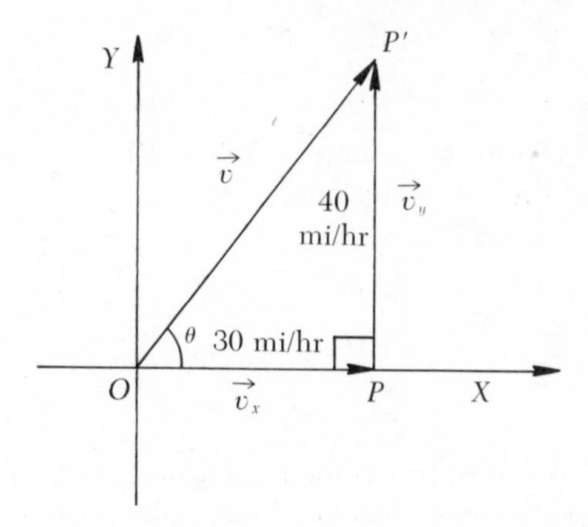

To determine the magnitude of $\vec{v}$ we note that $\angle OPP' = 90°$. Making use of the Pythagorean theorem gives

$$v = \sqrt{v_x^2 + v_y^2}$$
$$= \sqrt{(30)^2 + (40)^2} \text{ mi/hr} = \sqrt{2500} \text{ mi/hr} = 50 \text{ mi/hr}$$

To determine the angle θ we make use of the definition of $\cos\theta$. From the figure,

$$\cos\theta = \frac{OP}{OP'} = \frac{v_x}{v}$$

Substituting $v_x = 30$ mi/hr and $v = 50$ mi/hr gives

$$\cos\theta = \frac{30 \text{ mi/hr}}{50 \text{ mi/hr}} = \frac{30}{50} = 0.60$$

From the Trigonometric Tables on page 99, we obtain

$$\theta = 53°$$

Therefore, the vector $\vec{v}$ is a velocity of 50 mi/hr directed at an angle of 53° with respect to the positive X-axis.

Exercises

The vector $\vec{F}$ represents a force of 500 N directed along the positive X-axis. Determine the following quantities:

1. $F =$ ____ (Ans. 154)
2. $\theta =$ ____ (Ans. 155)

3. $F_x =$ ____ (Ans. 156)

4. $F_y =$ ____ (Ans. 157)

The vector $\vec{F}$ represents a force of 500 N directed along the positive Y-axis. Determine the following quantities:

5. $F =$ ____ (Ans. 158)

6. $\theta =$ ____ (Ans. 159)

7. $F_x =$ ____ (Ans. 160)

8. $F_y =$ ____ (Ans. 161)

The vector $\vec{v}$ represents a velocity of 10 m/s directed at an angle of 20° with respect to the positive X-axis. Determine the following quantities:

9. $v =$ ____ (Ans. 162)

10. $\theta =$ ____ (Ans. 163)

11. $v_x =$ ____ (Ans. 164)

12. $v_y =$ ____ (Ans. 165)

The vector $\vec{F}$ represents a force of 20 N directed at an angle of 45° with respect to the positive X-axis. Determine the following quantities:

13. $F =$ ____ (Ans. 166)

14. $\theta =$ ____ (Ans. 167)

15. $F_x =$ ____ (Ans. 168)

16. $F_y =$ ____ (Ans. 169)

The rectangular coordinates of a force vector $\vec{F}$ are $F_x = 10$ N and $F_y = 20$ N. Determine the following quantities:

17. $F =$ ____ (Ans. 170)

18. $\theta =$ ____ (Ans. 171)

CHAPTER SIX

FURTHER EXAMPLES FROM THE PHYSICAL SCIENCES

In this chapter we discuss some additional problems from the various physical sciences. These examples have been chosen to illustrate and to extend some of the principles and techniques described in the previous chapters. In reading these examples you will see how simple mathematical ideas are used to solve real problems in the sciences.

6.1 AN ANCIENT MEASUREMENT OF THE EARTH'S CIRCUMFERENCE

The idea that the Earth is round was not a new thought in Columbus' time. In the days of the early Greeks (several centuries before Christ) the world was generally considered to be round. The first recorded determination of the Earth's circumference was made by Eratosthenes (about 276–196 B.C.), a Greek astronomer and director of the library at Alexandria.

Eratosthenes was well schooled in the science and mathematics of his time, and he made measurements based on sound astronomical reasoning. He observed that at noon on a particular day (according to the modern calendar the date was June 21, the *summer solstice* or first day of summer), the Sun's rays cast no shadow on the floor of a deep, vertical well in the town of Syene, Egypt (modern Aswan). That is, the noon Sun stood directly overhead at Syene so that the rays were exactly parallel to the sides of the well shaft and produced no shadow. Eratosthenes found that at noon on the same day, the Sun did not stand directly overhead at Alexandria, a distance to the north of Syene. He measured the direction of the noon rays at Alexandria and discovered that they made an angle with the vertical equal to 1/50 of a complete circle (that is, $360°/50 = 7.2°$, as shown in Fig. 6.1). Eratosthenes reasoned that if there is an angle of 7.2° between the directions of the Sun's rays at Syene and Alexandria,

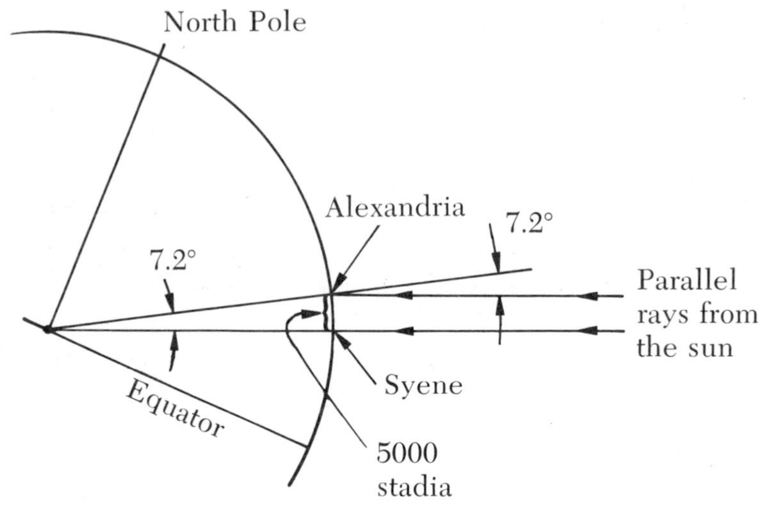

FIGURE 6.1 Eratosthenes' method for determining the circumference of the Earth. Because the Sun's rays are parallel, the angle that the rays make with the local vertical at Alexandria is equal to the angle between the lines connecting Syene and Alexandria with the center of the Earth.

there must also be an angle of 7.2° between the lines connecting these points with the center of the Earth. That is, the distance from Syene to Alexandria must be 1/50 of the total distance around the Earth. Next, Eratosthenes measured the distance between the two cities, obtaining a value of 5000 *stadia*. (The *stade*—or *stadium*—is an ancient unit of length; the stade used by Eratosthenes is believed to be equal to 607 ft.) He then calculated the circumference of the Earth to be $50 \times 5000 =$ 250 000 stadia.

Eratosthenes' reasoning was certainly correct. But he did make the assumption, not proven at the time, that the Sun is sufficiently far away for the light rays to be parallel at the position of the Earth. (Can you see why this assumption is a necessary part of the argument?) The problem with interpreting Eratosthenes' result is that we do not know with certainty the size of a stade in modern units. If we take 1 stade = 607 ft, we obtain a value of 28 750 mi for Eratosthenes' value of the Earth's circumference. This is reasonably close to the modern value of 24 920 mi or 40 090 km (at the Equator).

We usually consider the Earth to be perfectly round, except for the irregular surface features, which are quite minor in a comparison with the size of the Earth. (The height of Mt. Everest is less than 1/1000th of the Earth's diameter.) But the actual shape of the Earth is not exactly spherical. Because the Earth is not absolutely rigid, and because it spins on a north-south axis, the Earth has developed a bulge at the Equator and it has become flattened at the poles. The amount of bulging and flattening is not great—the equatorial radius is only about 14 mi larger

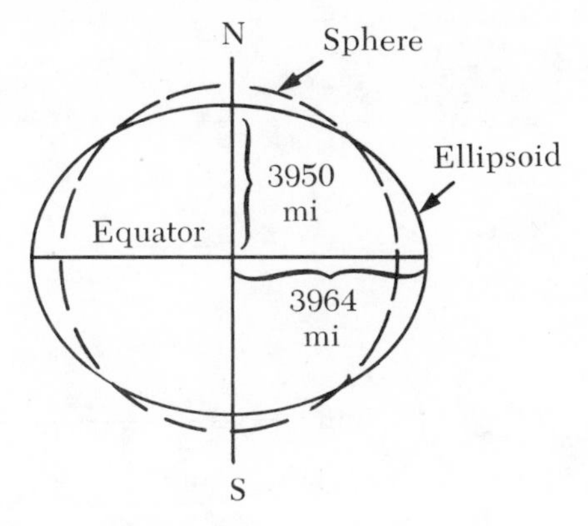

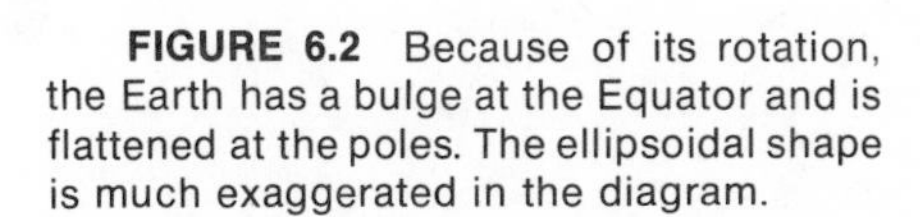
FIGURE 6.2 Because of its rotation, the Earth has a bulge at the Equator and is flattened at the poles. The ellipsoidal shape is much exaggerated in the diagram.

than the polar radius (see Fig. 6.2). Modern measurements have shown the dimensions of the Earth to be:

$$\begin{aligned} \text{Equatorial radius of the Earth} &= 3964 \text{ mi} \\ &= 6378 \text{ km} \\ \text{Polar radius of the Earth} &= 3950 \text{ mi} \\ &= 6356 \text{ km} \end{aligned} \tag{6.1}$$

We usually take "the" radius of the Earth to be 6.38×10^6 m, or approximately 4000 mi.

6.2 THE MEASUREMENT OF PLANETARY DISTANCES

In the early days when astronomers were struggling with the problem of planetary motion, their data were derived solely from observations of the positions of planets in the sky. But they had no established distance scale, no planetary measuring-stick, by which they could determine the distances to the planets. Except for relatively crude estimates, it was possible only to express all distances in the solar system in terms of the Earth-Sun distance, which is called the *astronomical unit* (A.U.).

To see how this method works, consider the case of the planet Mars. The time required for Mars to complete one revolution around the Sun (that is, the Martian year) is 687 days. Therefore, on two dates, 687 days apart, Mars will be in the same position in its orbit but the Earth will be in different positions (Fig. 6.3). The Earth year is 365 days and so, during one Martian year, the Earth will execute one complete revolution around the Sun and will be $2 \times 365 - 687 = 43$ days short of completing the second orbit. In 365 days the Earth moves through 360°; therefore, 43 days corresponds to an angular motion of $43 \times (360°/365) = 42°$.

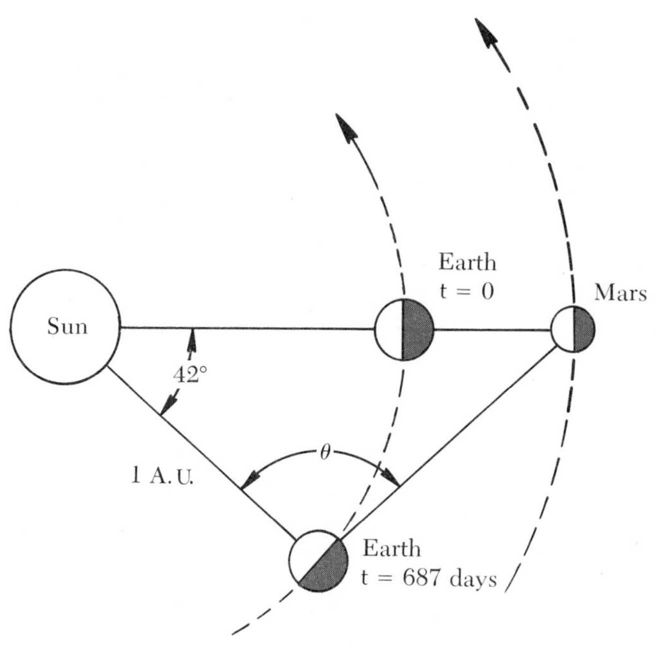

FIGURE 6.3 Geometry for the determination of the Mars-Sun distance in terms of the Earth-Sun distance (the *astronomical unit*).

Consequently, when Mars returns to its original position, the Earth-Sun line on that day will make an angle of 42° with the starting Earth-Sun line. Figure 6.3 shows the geometry of the situation for the initial condition (time $t = 0$) in which Mars is at *opposition* (that is, the Sun, Earth, and Mars are in line). The angle θ, which specifies the position of Mars in the sky, can be measured directly at $t = 687$ days; the result is $\theta = 97°$. The Sun-Earth-Mars triangle is now determined—two angles and one side (the Earth-Sun distance = 1 A.U.) are known. Therefore, the length of the side that is the Mars-Sun distance can be computed in terms of the astronomical unit; the result is that the Mars-Sun distance is 1.524 A.U. Similar geometric (or trigonometric) techniques can be used to determine the distances to the other planets in A.U.

But what is the length of the astronomical unit in miles or in meters? The best method that we have available to make such a measurement involves a technique called *radar-ranging.* If a pulsed radio wave is emitted by a radar transmitter, it will travel outward with a speed equal to the speed of light, 186 000 miles per second or 3×10^8 meters per second. When the radar wave strikes a distant object, a portion of the wave will be reflected and will travel back to the receiver where it is detected. The time interval between transmission and reception of the radar signal, together with the known speed of the pulse, allows a determination of the distance to the object. Suppose that the planet Venus is in the position shown in Figure 6.4. The angle ϕ_1 is measured directly and the angle ϕ_2 is determined from the orbit char-

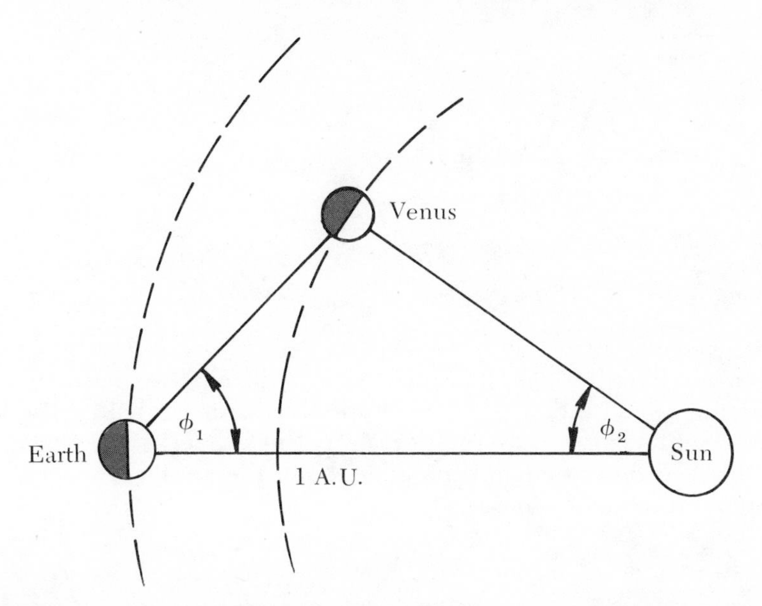

FIGURE 6.4 By bouncing radar waves off Venus, the Earth-Venus distance is measured. This result, combined with a knowledge of the angles ϕ_1 and ϕ_2 permits a determination of the astronomical unit in meters.

acteristics by a method similar to that described above for the case of Mars. The transit time for a radar signal to travel to Venus and return is measured, and the Earth-Venus distance is calculated. Again, we have a triangle with two angles and one side known. Therefore, the side which is the Earth-Sun distance can be determined in miles or meters. But this distance is just the *astronomical unit.* Modern measurements have shown that

$$\begin{aligned} 1 \text{ A.U.} &= 92\ 955\ 700 \text{ miles} \\ &= 1.495\ 979 \times 10^{11} \text{ meters} \end{aligned} \tag{6.2}$$

For most purposes in the physical sciences we usually use the approximate values of 93 million miles and 1.5×10^{11} meters.

The value of the astronomical unit determined by radar-ranging is the standard length for *all* astronomical distance measurements.

6.3 RADIAN MEASURE AND THE SUN'S DIAMETER

As we discussed in Section 3.1, the most commonly used unit of angular measure is the *degree,* which is 1/360 of a complete circle. For many types of problems in the physical sciences it proves more convenient to use another unit called the *radian* (rad). If we measure the length of arc along the circumference of a circle (see Fig. 6.5), we find

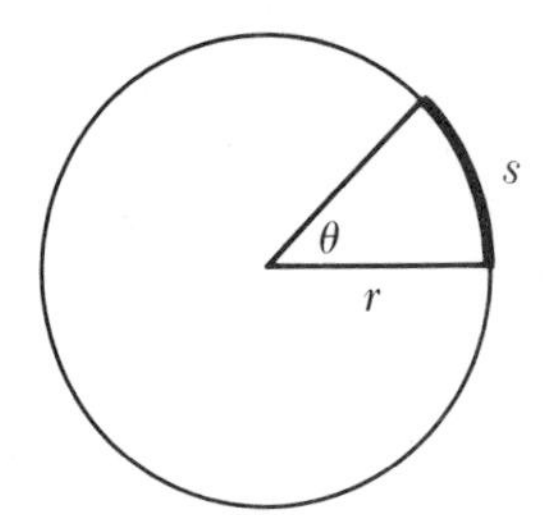

FIGURE 6.5 If $s = r$, the angle θ is equal to 1 radian.

that the arc length s is proportional to the angle θ between the two radii that define the arc; that is, $s \propto \theta$. Furthermore, if we hold θ fixed and increase r, then s increases in direct proportion; that is, $s \propto r$. *One radian* is defined to be the angle subtended when the arc length s is exactly equal to the radius r. Thus,

$$s = r\theta \tag{6.3}$$

where θ is measured in radians.

If θ is increased until it is equal to 360°, the arc s is just the circumference, $2\pi r$. Then, $s = 2\pi r = r\theta$, so that $\theta = 2\pi$ radians corresponds to $\theta = 360°$. Therefore,

$$1 \text{ rad} = \frac{360°}{2\pi} = 57.2958° \cdots \cong 57.3° \tag{6.4}$$

In order to find the radian equivalent of 1°, we write

$$1° = \frac{2\pi}{360°} = 0.01745 \cdots \text{ rad} \tag{6.5}$$

Notice that although the *radian* is a unit of angular measure, it does not have physical dimensions. Therefore, when we put values into Equation 6.3, such as 3 cm = (1 rad) × (3 cm), we do not have different physical dimensions on the two sides of the equation. Usually, we carry the designation *radian* in our equations as a reminder of the angular units we are using, but when the final answer is obtained, it is sufficient to include only the *physical* dimensions. For example, if we calculate the length of the arc of a circle with $r = 10$ cm that is intercepted by an angle $\theta = 0.5$ radian, we obtain

$$s = r\theta = (10 \text{ cm}) \times (0.5 \text{ rad}) = 5 \text{ cm}$$

The fact that $s = r\theta$ provides a method for closely estimating distances in certain circumstances. Suppose that a pole is placed vertically in the ground a distance R away from an observer located at a point O (Fig. 6.6). The observer measures the angle subtended by the pole and

FIGURE 6.6 Estimating the height of an object by measuring the angle subtended at O; $h \cong R\theta$.

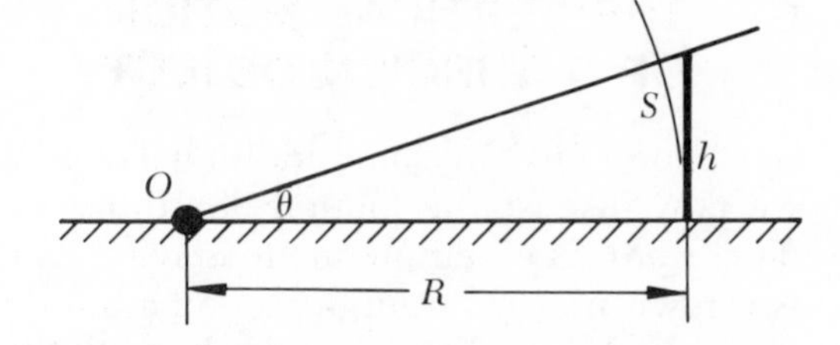

finds it to be θ. Mentally, the observer constructs a circle of radius R centered at O and passing through the bottom of the pole. The arc length S is

$$S = R\theta \tag{6.6}$$

Since θ is a small angle, S is approximately equal to the height of the pole h, and we can write

$$h \cong R\theta \tag{6.7}$$

This method for obtaining approximate values for h will be useful if θ is sufficiently small. Even for θ as large as 20°, the error is only about 4 per cent. If θ is a few degrees, the error is usually negligible for most purposes; for $\theta = 1°$, the error is 0.01 per cent or 1 part in 10^4. (Remember that when using Equation 6.6 or 6.7 the angle θ must be expressed in *radians*.)

In the previous section we learned that the Earth-Sun distance is 93 million miles (93×10^6 mi) or 1.5×10^{11} meters. If we sight toward the Sun and imagine lines drawn from the eye to opposite ends of a diameter of the Sun's disk, we find that the angle between the two lines is approximately one-half degree—actually, 0.53°. This angle, together with the Earth-Sun distance is sufficient to determine the Sun's diameter.

First, we must convert 0.53° to radians:

$$\begin{aligned} 0.53° &= (0.53°) \times (0.01745 \text{ rad/deg}) \\ &= 0.009\,245 \text{ rad} \end{aligned}$$

Next, we use Equation 6.6 and write for the Sun's diameter d,

$$\begin{aligned} d &= R\,\theta \\ &= (1.5 \times 10^{11} \text{ m}) \times (0.009\,245 \text{ rad}) \\ &= 1.39 \times 10^9 \text{ m} \end{aligned}$$

or, approximately 860 000 mi.

6.4 THE VERTICAL MOTION OF A THROWN OBJECT

In order to gain some additional familiarity with quadratic equations, we now discuss the motion of a thrown object which experiences the downward acceleration due to gravity. As illustrated in Figure 6.7, a ball is thrown upward from the top of a building. The initial upward speed of the ball is 19.6 m/s and the height of the building is 49 m. We wish to find

(a) the time required for the ball to reach maximum height above the top of the building,
(b) the maximum height to which the ball rises, and
(c) the time required for the ball to strike the pavement.

Figure 6.7 shows that we choose the top of the building to correspond to the position $x = 0$. We also choose the *positive* direction of x to be downward. Therefore, the initial upward motion of the ball is in the *negative* X-direction: $v_0 = -19.6$ m/s. The downward acceleration due to gravity is in the positive X-direction: $g = 9.8$ m/s^2.

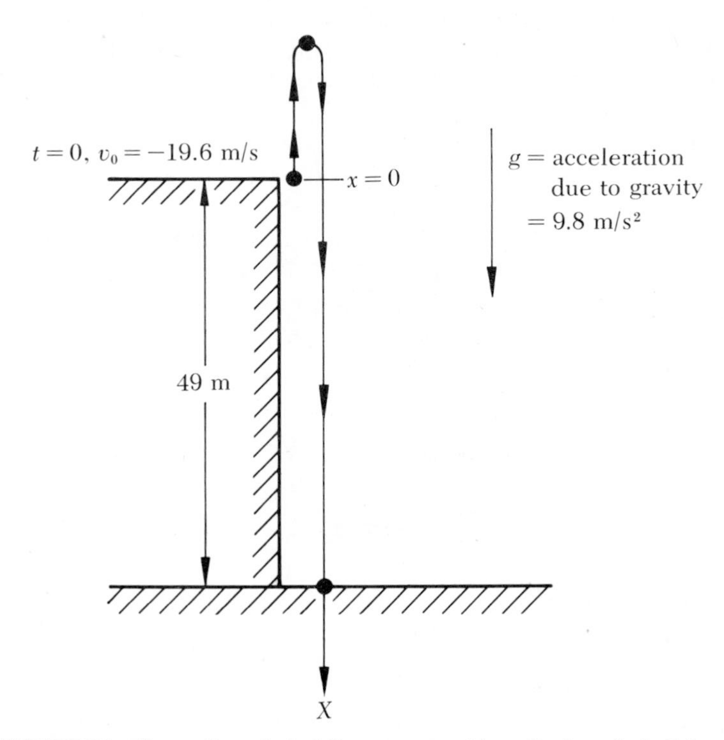

FIGURE 6.7 The motion of a ball thrown upward from the top of a building.

In the study of the kinematics of moving object, we learn that the velocity of an object that experiences an acceleration a is given by

$$v = v_0 + at \tag{6.8}$$

where v_0 is the velocity at the instant $t = 0$. Furthermore, we find that the displacement of the object (which starts at $x = 0$ at $t = 0$) is given by

$$x = v_0 t + \frac{1}{2} at^2 \tag{6.9}$$

In this problem we have

$$\left.\begin{aligned} v_0 &= -19.6 \text{ m/s} \\ a = g &= 9.8 \text{ m/s}^2 \end{aligned}\right\} \tag{6.10}$$

Substituting these values into Equations 6.8 and 6.9, we obtain

$$v = -19.6 + 9.8\ t \qquad (MKS \text{ units: } v \text{ in m/s and } t \text{ in s}) \tag{6.11}$$

$$x = -19.6\ t + 4.9\ t^2 \qquad (MKS \text{ units: } x \text{ in m and } t \text{ in s}) \tag{6.12}$$

In order to answer part (a), we note that at its maximum height above the building, the ball has zero velocity. Therefore, the time t at which this occurs is found by setting $v = 0$ in Equation 6.11:

$$0 = -19.6 + 9.8\ t$$

Solving for t, we find

$$t = \frac{19.6}{9.8} = 2 \text{ s}$$

Thus, the ball reaches its maximum height 2 seconds after it is released.

In answer to part (b), to determine the maximum height to which the ball rises, we substitute $t = 2$ s into Equation 6.12 and solve for x. This gives

$$\begin{aligned} x &= -19.6 \times 2 + 4.9 \times (2)^2 \\ &= -39.2 + 19.6 \\ &= -19.6 \text{ m} \end{aligned}$$

Therefore, we conclude that the ball rises to a maximum height of 19.6 m above the top of the building. (It should be kept in mind that *negative* values of x correspond to displacements *above* the top of the building.)

In answer to part (c), when the ball strikes the pavement, its *net displacement* is $x = 49$ m. Making use of Equation 6.12, the time t at which this occurs is determined from

$$49 = -19.6\ t + 4.9\ t^2$$

Rearranging terms and dividing by 4.9, we find

$$t^2 - 4\ t - 10 = 0$$

This is a *quadratic* equation for the unknown quantity t. The solution is given by (see Section 2.3)

$$t = \frac{4 \pm \sqrt{4^2 + 40}}{2} = 2 \pm \sqrt{14}$$

The two solutions are

$$t_+ = 2 + \sqrt{14} = 2 + 3.74 = 5.74\ \text{s}$$
$$t_- = 2 - \sqrt{14} = 2 - 3.74 = -1.74\ \text{s}$$

The ball was released at $t = 0$; therefore, the solution $t_- = -1.74$ s, has no meaning in this case. We conclude that the ball strikes the ground 5.74 seconds after it is released.

6.5 A PROBLEM IN KINEMATICS

As illustrated in Figure 6.8, a block of mass $m = 1$ kg is initially at rest ($v_0 = 0$) on a horizontal frictionless surface. A constant force $F = 10$ N is applied to the block in the positive X-direction. We wish to find

(a) the acceleration of the block,
(b) the velocity of the block after 10 s, and
(c) the displacement of the block after 10 s.

In answer to part (a), we make use of Newton's Second Law,

$$\text{Force} = (\text{mass}) \times (\text{acceleration}) \tag{6.13}$$

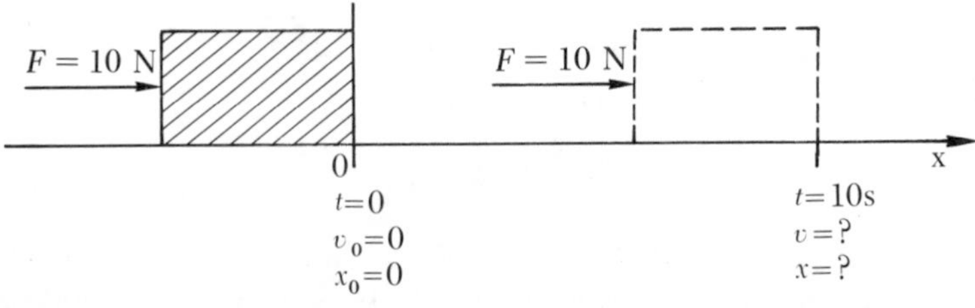

FIGURE 6.8 A constant force $F = 10$ N is applied to a block initially at rest on a horizontal frictionless surface.

That is,

$$F = ma \tag{6.14}$$

where F, m, and a denote the force, mass and acceleration, respectively. Solving Equation 6.14 for the acceleration a, and making use of $m = 1$ kg and $F = 10$ N $= 10$ kg-m/s², we find

$$a = \frac{F}{m} = \frac{10 \cancel{\text{kg}}\text{-}\frac{m}{s^2}}{1 \cancel{\text{kg}}} = 10 \frac{m}{s^2}$$

In answer to part (b), we recall that the velocity v can be expressed as

$$v = v_0 + at \tag{6.15}$$

where v_0 is the initial $(t = 0)$ velocity. Because $a = 10$ m/s², and $v_0 = 0$ (the block is initially at rest), Equation 6.15 reduces to

$$v = 0 + 10\ t$$

That is,

$$v = 10\ t \text{ (}MKS\text{ units: } v \text{ in m/s and } t \text{ in s)}$$

Substituting $t = 10$ s, we find

$$v = 10 \times 10 = 100 \frac{m}{s}$$

Therefore, after 10 s, the block is traveling with velocity $v = 100$ m/s.

In answer to part (c), we recall that the displacement x can be expressed as

$$x = x_0 + v_0 t + \frac{1}{2} at^2 \tag{6.16}$$

where x_0 is the initial $(t = 0)$ displacement. Without loss of generality, we choose the origin O such that (see the figure)

$$x_0 = 0$$

Substituting $x_0 = 0$, $v_0 = 0$ and $a = 10$ m/s² in Equation 6.16, we find

$$x = 0 + 0 + \frac{1}{2} \times 10\ t^2$$

That is,

$$x = 5\ t^2 \text{ (}MKS\text{ units: } x \text{ in m and } t \text{ in s)}$$

Substituting $t = 10$ s gives

$$x = 5 \times (10)^2 = 500 \text{ m}$$

Therefore, after 10 s, the block has been displaced 500 m.

6.6 THE IDEAL GAS LAW—EXPANSION OF A RISING BALLOON

If you have ever witnessed or seen photographs of the launching of a high-altitude balloon, you may have noticed that the gas bag is floppy and only partially filled at the time of release. Why is the bag not filled? Would not this provide greater lift and a more rapid ascent? To answer these questions, we must inquire into the behavior of gases under various physical conditions.

The readily measurable bulk properties of a gas are the *mass*, the *temperature*, the *volume*, and the *pressure*. Experiments have shown that these properties of gases (for temperatures above the liquefaction points) are related in a particularly simple way. For a given mass of gas, the temperature T, the pressure P, and the volume V are connected by the following relationship:

$$\frac{PV}{T} = \text{constant} \qquad (6.17)$$

This equation states that no matter how the pressure, volume, and temperature are varied for a particular sample of gas, the quantity PV/T remains constant. For different samples of gas, the value of the constant will be different but the connection among P, V, and T will be the same.

Real gases follow Equation 6.17 closely but not *exactly*. Nevertheless, the equation is sufficiently accurate for most purposes. We say that an *ideal* gas would obey Equation 6.17 exactly; therefore, this relationship is called the *ideal gas law*.

One important point to notice regarding the ideal gas law equation concerns the temperature. If we maintain the pressure of a gas sample at a constant value, then the equation can be expressed more simply as $V/T = \text{constant}$, or $V \propto T$. That is, at constant pressure, the volume of a gas sample is directly proportional to the temperature. What scale do we use to measure temperature? The *Fahrenheit* scale is most often used in the U.S. for everyday temperatures. But the *centigrade* or *Celsius* scale is the most widely used scale for scientific purposes. However, neither of these scales is appropriate for use in the ideal gas law equation. The reason is easy to see. Suppose that we begin with a gas sample that has a volume of 10 m^3 at $T = 20°$ C. What will be the volume if T is lowered to $-20°$ C (P remaining constant)? The proportionality $V \propto T$ indicates that the volume would be -10 m^3. But a negative volume has no physical meaning. We must therefore use a temperature scale for which T never becomes negative. We can ensure this if we choose the lowest possible or imaginable temperature as the zero point of the scale; this point is called *absolute zero.* The temperature scale so constructed is called the *absolute* scale or *Kelvin* scale. The size of the degree on this scale is the same as that on the centigrade scale, and temperatures on this scale are abbreviated °K. The zero point of the absolute scale corresponds to $-273°$ C; that is, $0°K = -273°$ C. In general,

$$T = T_C + 273° \qquad (6.18)$$

where T is the temperature on the absolute scale and where T_C is the temperature on the centigrade scale.

How do we use the ideal gas law equation? Suppose that we begin with a set of conditions labeled P_1, V_1, and T_1. Then, we can write

$$\frac{P_1 V_1}{T_1} = \text{constant} \qquad (6.19)$$

After any changes have taken place, we have a new set of conditions, P_2, V_2, and T_2. We write

$$\frac{P_2 V_2}{T_2} = \text{constant} \qquad (6.20)$$

Because the *constant* appearing in the two equations above has the same value, we can combine the equations as

$$\frac{P_1 V_1}{T_1} = \frac{P_2 V_2}{T_2} \qquad (6.21)$$

Then, if we know five of the six values that appear in the expression, we can calculate the value of the remaining quantity.

Let us return to the example of the high-altitude balloon. The initial pressure is sea-level atmospheric pressure; let us call this P_0. The initial temperature will be, for example, 15° C (or 59° F); using Equation 6.18, $T_1 = 15° + 273° = 288°$ K. A typical high-altitude research balloon designated to carry a 1000-pound load of equipment to an altitude of 100 000 feet will be filled (at sea level) with about 400 m³ of helium; therefore, we use $V_1 = 400$ m³. At the final altitude of 100 000 ft, the temperature will be about $-50°$ C, or $T_2 = -50° + 273° = 223°$ K. The pressure at this altitude is only about 1/100 of sea-level pressure; therefore, $P_2 = 0.01\ P_0$. Summarizing,

$$P_1 = P_0;\ T_1 = 288°\ \text{K};\ V_1 = 400\ \text{m}^3$$

$$P_2 = 0.01\ P_0;\ T_2 = 223°\ \text{K};\ V_2 = ?$$

Solving Equation 6.21 for V_2, we find

$$V_2 = \frac{P_1}{P_2} \times \frac{T_2}{T_1} \times V_1$$

Substituting,

$$V_2 = \frac{P_0}{0.01\,P_0} \times \frac{223^\circ\,\text{K}}{288^\circ\,\text{K}} \times 400\ \text{m}^3$$

$$= 31\,000\ \text{m}^3$$

or, approximately 77 times the original volume.

We can now understand why the balloon is filled initially to only a small fraction of its capacity. It is true that a greater filling would result in a faster initial rise, but the expansion with altitude would quickly fill the bag to capacity and cause a rupture before the balloon reached the desired altitude.

In computing V_2, notice that we did not need to express the pressure values in the ordinary units of pressure (newtons per square meter). Because the *ratio*, P_1/P_2, is involved, we need only to express the two pressures in a common way. Here, we wrote both pressures in terms of sea-level pressure, P_0.

6.7 A PROBLEM IN CHEMICAL SOLUBILITY

If you add some common table salt (sodium chloride, NaCl) to a glass of water and stir the mixture, you will find that the salt disappears. The salt dissolves and a *solution* is formed. If you continue to add salt, you will discover that at a certain point no more salt will dissolve—the solution is *saturated.* Once the condition of saturation is reached, any additional salt that is placed in the water will simply settle to the bottom and will remain as solid salt. If you have 100 g of water at room temperature (20°C), the maximum amount of sodium chloride that will go into solution is approximately 36 g. Other chemical salts, such as potassium nitrate (KNO_3), ammonium iodide (NH_4I), and calcium chloride ($CaCl_2$), exhibit the same property of saturation in solution.

The *solubility* of a substance in water is usually given in terms of the number of grams of the substance that will produce a saturated solution in 100 g of water. The solubility of most chemical salts increases with temperature, some slowly, and some quite rapidly.

Suppose that we have 1/2 liter (500 cm³) of water at a temperature of 75°C. We add potassium nitrate (KNO_3) until the solution is saturated. Next, we cool the water to 18°C. How much KNO_3 remains in solution and how much will form in the solid state on the bottom of the container?

We need to know the solubility of KNO_3 in water at 75°C and at 18° C. If we look in a collection of tabulated chemical data (for example, *The Handbook of Chemistry and Physics*), we find the solubility values

TABLE 6.1 SOLUBILITY OF POTASSIUM NITRATE (KNO_3)

Temperature (°C)	Solubility (grams in 100 grams of water)
0°	13.3
10°	20.9
20°	31.6
30°	45.8
40°	63.9
50°	85.5
60°	110.0
70°	138
80°	169
90°	202
100°	246

listed in Table 6.1. This table does not include entries for the two temperatures desired, 75°C and 18°C. Therefore, we make a graph of the data by plotting the solubility values versus temperature. This graph is shown in Figure 6.9, in which we have connected the data points by a smooth curve. From the curve we read the required values:

Solubility of KNO_3 at 75°C = 153 g in 100 g of water

Solubility of KNO_3 at 18°C = 28 g in 100 g of water

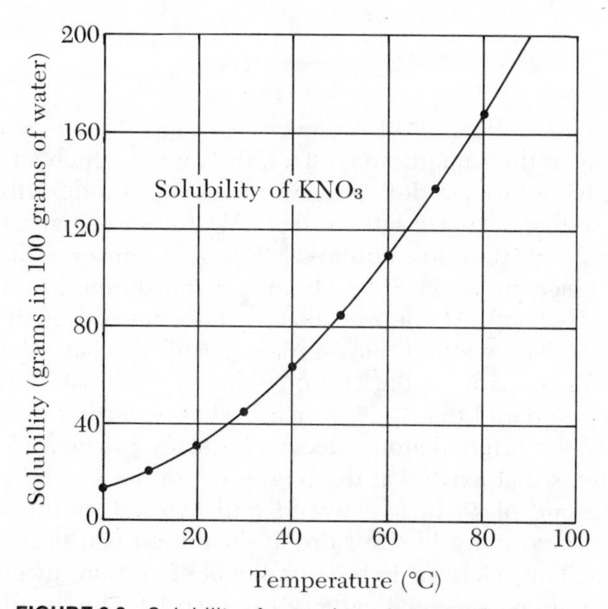

FIGURE 6.9 Solubility of potassium nitrate (KNO_3) in water.

The container holds 500 cm^3 of water. Because the density of water is 1 g/cm^3, the mass of the water is 500 g. Therefore, at the two temperatures, the amounts of KNO_3 in solution are

$$\text{at } 75°C: \quad 5 \times 153 \text{ g} = 765 \text{ g}$$

$$\text{at } 18°C: \quad 5 \times 28 \text{ g} = 140 \text{ g}$$

Therefore, when the saturated solution at 75°C is cooled to 18°C, 140 g of KNO_3 remains in solution and 765 g − 140 g = 625 g will form as solid KNO_3 on the bottom of the container.

6.8 RADIOACTIVE DECAY

A *radioactive* material is a substance whose nuclei spontaneously emit a subatomic particle (either an electron or the nucleus of a helium atom) and are thereby transformed into a different atomic species. For example, radium-226 (^{226}Ra) emits a helium nucleus (which is called an *α particle*) and becomes radon-222 (^{222}Rn); this process is called α decay:

$$^{226}\text{Ra} \xrightarrow{\alpha \text{ decay}} {}^{222}\text{Rn}$$

Also, when a nucleus of carbon-14 (^{14}C) undergoes *β decay* by emitting an electron, a nucleus of nitrogen-14 (^{14}N) is formed:

$$^{14}\text{C} \xrightarrow{\beta \text{ decay}} {}^{14}\text{N}$$

If we have a collection of radioactive atoms, the atoms do not all undergo decay at the same time. In fact, the rate at which a radioactive material decays follows a definite law which can be described as follows. Suppose that at time $t = 0$ we have N_0 atoms of a particular radioactive species, for example, sodium-24 (^{24}Na). We observe that β decay events take place in which ^{24}Na atoms are transformed into atoms of magnesium-24 (^{24}Mg). At the end of a certain interval of time (15 hr, for the case of ^{24}Na), we find that $\frac{1}{2} N_0$ atoms of ^{24}Na have decayed and that $\frac{1}{2} N_0$ atoms remain. At the end of 30 hr, we find that $\frac{3}{4} N_0$ atoms of ^{24}Na have decayed and that $\frac{1}{4} N_0$ remain. That is, during the first 15-hr period, half of the original atoms decayed; during the next 15-hr period half of the atoms that existed at the beginning of the second period decayed. At the end of 45 hr, we would find that half of the atoms that existed at the beginning of the third 15-hr period had decayed; $\frac{1}{8} N_0$ would remain. This decrease in the number of ^{24}Na atoms by half in each 15-hr period is shown schematically in Figure 6.10. The time interval of 15 hr for the decay of half the number of ^{24}Na atoms is called the *half-*

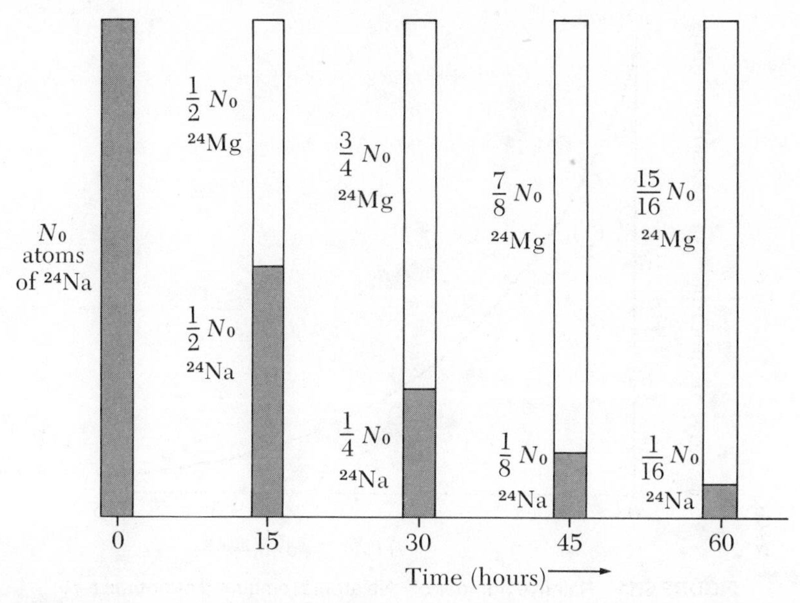

FIGURE 6.10 In any collection of ^{24}Na atoms, one-half of the atoms that exist at any time will have decayed 15 hours after that time.

life. Half-lives for other radioactive species range from tiny fractions of a second to billions of years.

Figure 6.10 shows only the number of ^{24}Na atoms remaining at multiples of 15 hr from the initial time. If we plot the number of atoms remaining at *any* instant after $t=0$ we have the curve shown in Figure 6.11. This type of curve is called the *exponential decay curve* and can be represented by the following mathematical expression:

$$N = N_0 e^{-t/\tau} \tag{6.22}$$

where N_0 is number of atoms at time $t=0$ and N is number of atoms remaining at time t. The quantity e is a certain irrational number that has the approximate value, $e \cong 2.7183$. Finally, τ is related to the *half-life* $t_{1/2}$ according to

$$t_{1/2} = 0.693\tau \tag{6.23}$$

In Equation 6.23, τ is called the *mean lifetime* of the substance.

To see how Equation 6.22 works, let us substitute into the expression a time t equal to two half-lives; that is,

$$t = 2t_{1/2} = 2 \times 0.693\,\tau = 1.386\,\tau$$

Then,

$$N = N_0 e^{-1.386}$$

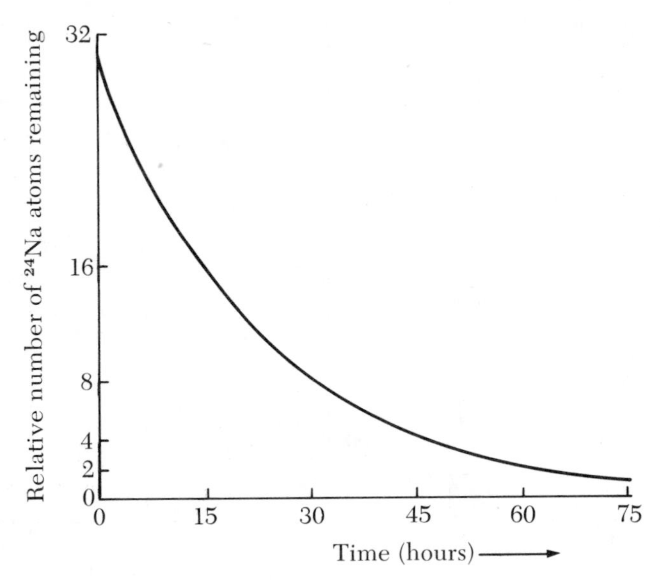

FIGURE 6.11 Relative number of ^{24}Na atoms remaining after time $t = 0$.

The evaluation of a number such as $e^{-1.386}$ requires special methods which we will not develop here. However, Table 6.2 lists values of e^{-x} for several values of x. We can see that $e^{-1.386}$ is close to $e^{-1.4} = 0.247$. In fact, $e^{-1.386} = 0.25$. Thus, Equation 6.23 shows that at $t = 2\,t_{1/2}$, $N = 1/4\ N_0$; this is exactly the result we expect based on our previous discussion.

The curve in Figure 6.11, which can be plotted accurately by using the values in Table 6.2, decreases smoothly toward zero (but never actually reaches zero). If we plot the same points again on a logarithmic graph (see Sect. 4.4), we obtain the results illustrated in Figure 6.12. We see that on a logarithmic scale the curve representing the number of radioactive atoms remaining at time t is a straight line. In this kind of graph the vertical scale remains compact even though the quantity plotted varies by several factors of 10.

TABLE 6.2 VALUES OF e^{-x}

x	e^{-x}	x	e^{-x}	x	e^{-x}
0	1.000	1.2	0.301	2.4	0.091
0.2	0.819	1.4	0.247	2.6	0.074
0.4	0.670	1.6	0.202	2.8	0.061
0.6	0.549	1.8	0.165	3.0	0.050
0.8	0.449	2.0	0.135	3.2	0.041
1.0	0.368	2.2	0.111	3.4	0.033

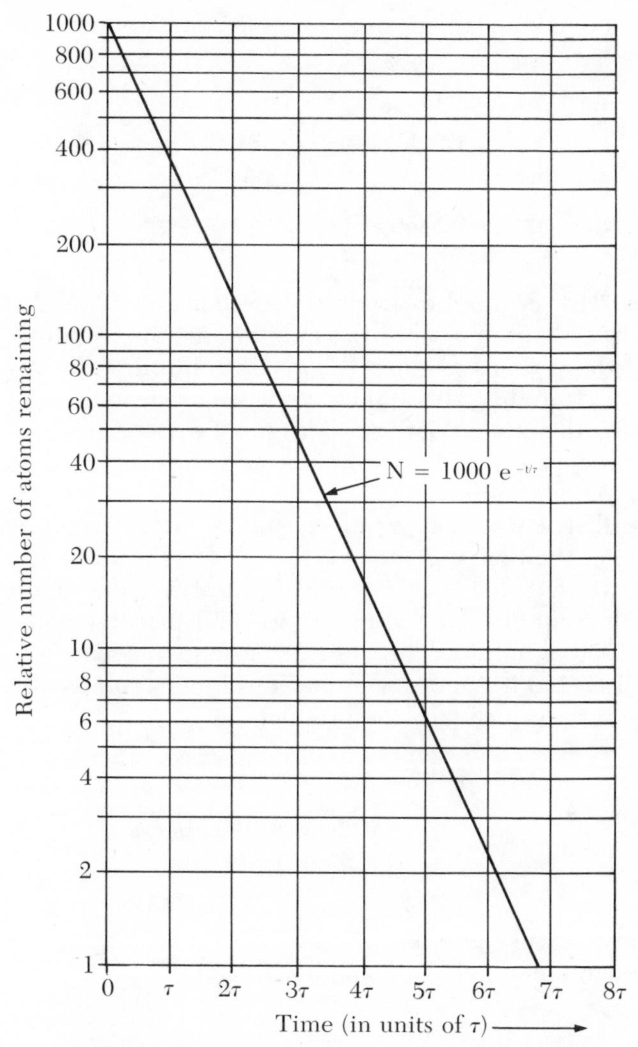

FIGURE 6.12 Logarithmic graph of radioactive decay.

Suppose that we have a sample of radioactive phosphorus-33 (^{33}P); the half-life of this substance is 25 days. At time $t = 0$ the sample is placed in a position near a detector and the apparatus registers decay events at a rate of 1000 per minute. At what rate will decay events be registered 65 days later and 166 days later?

First, we must compute the mean lifetime τ for ^{33}P. Using Equation 6.23, we find

$$\tau = \frac{t_{1/2}}{0.693} = \frac{25 \text{ days}}{0.693} = 36.08 \text{ days}$$

Next, we need to know the desired times (65 days and 166 days) in terms of τ:

$$65 \text{ days} = \frac{65}{36.08}\,\tau = 1.8\,\tau$$

$$166 \text{ days} = \frac{166}{36.08}\,\tau = 4.6\,\tau$$

That is, 65 days is equal to 1.8 mean lifetimes of ^{33}P, and 166 days is equal to 4.6 mean lifetimes of ^{33}P. To determine the counting rate of the detector at these times, we use Figure 6.12. Because the vertical scale of the decay graph indicates 1000 at $t = 0$, we can read the counting rates directly from this scale. Corresponding to 1.8 τ, we read the counting rate as approximately 170 per minute, and corresponding to 4.6 τ, we read 10 counts per minute.

Notice that if we wish greater accuracy in the result, we can use Table 6.2. For the time equal to 1.8 τ, we need the value of $e^{-1.8}$. From the table, we see that $e^{-1.8} = 0.165$. Multiplying this figure by 1000 counts per minute (the rate at $t = 0$), we find that the counting rate at 65 days is 165 per minute. This value is close to that read from the graph in Figure 6.12, but it is more accurate. To obtain a comparable value for 166 days (4.6 τ), we can take the product

$$\begin{aligned} e^{-4.6} &= e^{-2.0} \times e^{-2.6} \\ &= (0.135) \times (0.074) \\ &= 0.01 \end{aligned}$$

and multiplying this figure by 1000, we obtain a counting rate of 10 per minute, in agreement with the graphical value.

APPENDIX A

SOME USEFUL TABLES

TABLE I. TRIGONOMETRIC TABLES OF SINES AND COSINES
(From 0° to 90° in steps of 1°)

θ Degrees	sin θ	cos θ	θ Degrees	sin θ	cos θ	θ Degrees	sin θ	cos θ
0	0	1.000						
1	0.018	0.9998	31	0.52	0.86	61	0.87	0.48
2	0.035	0.999	32	0.53	0.85	62	0.88	0.47
3	0.052	0.999	33	0.54	0.84	63	0.89	0.45
4	0.070	0.998	34	0.56	0.83	64	0.90	0.44
5	0.087	0.996	35	0.57	0.82	65	0.91	0.42
6	0.10	0.995	36	0.59	0.81	66	0.91	0.41
7	0.12	0.993	37	0.60	0.80	67	0.92	0.39
8	0.14	0.990	38	0.62	0.79	68	0.93	0.37
9	0.16	0.988	39	0.63	0.78	69	0.93	0.36
10	0.17	0.985	40	0.64	0.77	70	0.94	0.34
11	0.19	0.98	41	0.66	0.75	71	0.95	0.33
12	0.21	0.98	42	0.67	0.74	72	0.95	0.31
13	0.23	0.97	43	0.68	0.73	73	0.96	0.29
14	0.24	0.97	44	0.69	0.72	74	0.96	0.28
15	0.26	0.97	45	0.71	0.71	75	0.97	0.26
16	0.28	0.96	46	0.72	0.69	76	0.97	0.24
17	0.29	0.96	47	0.73	0.68	77	0.97	0.23
18	0.31	0.95	48	0.74	0.67	78	0.98	0.21
19	0.33	0.95	49	0.75	0.66	79	0.98	0.19
20	0.34	0.94	50	0.77	0.64	80	0.985	0.17
21	0.36	0.93	51	0.78	0.63	81	0.988	0.16
22	0.37	0.93	52	0.79	0.62	82	0.990	0.14
23	0.39	0.92	53	0.80	0.60	83	0.993	0.12
24	0.41	0.91	54	0.81	0.59	84	0.995	0.10
25	0.42	0.91	55	0.82	0.57	85	0.996	0.087
26	0.44	0.90	56	0.83	0.56	86	0.998	0.070
27	0.45	0.89	57	0.84	0.54	87	0.999	0.052
28	0.47	0.88	58	0.85	0.53	88	0.999	0.035
29	0.48	0.87	59	0.86	0.52	89	0.9998	0.018
30	0.50	0.87	60	0.87	0.50	90	1.000	0

TABLE II. SQUARES AND ROOTS*

n	n^2	$\sqrt{n}$	$\sqrt{10n}$
1	1	1.000 000	3.162 278
2	4	1.414 214	4.472 136
3	9	1.732 051	5.477 226
4	16	2.000 000	6.324 555
5	25	2.236 068	7.071 068
6	36	2.449 490	7.745 967
7	49	2.645 751	8.366 600
8	64	2.828 427	8.944 272
9	81	3.000 000	9.486 833
10	100	3.162 278	10.00000
11	121	3.316 625	10.48809
12	144	3.464 102	10.95445
13	169	3.605 551	11.40175
14	196	3.741 657	11.83216
15	225	3.872 983	12.24745
16	256	4.000 000	12.64911
17	289	4.123 106	13.03840
18	324	4.242 641	13.41641
19	361	4.358 899	13.78405
20	400	4.472 136	14.14214
21	441	4.582 576	14.49138
22	484	4.690 416	14.83240
23	529	4.795 832	15.16575
24	576	4.898 979	15.49193
25	625	5.000 000	15.81139
26	676	5.099 020	16.12452
27	729	5.196 152	16.43168
28	784	5.291 503	16.73320
29	841	5.385 165	17.02939
30	900	5.477 226	17.32051
31	961	5.567 764	17.60682
32	1 024	5.656 854	17.88854
33	1 089	5.744 563	18.16590
34	1 156	5.830 952	18.43909
35	1 225	5.916 080	18.70829
36	1 296	6.000 000	18.97367
37	1 369	6.082 763	19.23538
38	1 444	6.164 414	19.49359
39	1 521	6.244 998	19.74842
40	1 600	6.324 555	20.00000
41	1 681	6.403 124	20.24846
42	1 764	6.480 741	20.49390
43	1 849	6.557 439	20.73644
44	1 936	6.633 250	20.97618
45	2 025	6.708 204	21.21320
46	2 116	6.782 330	21.44761
47	2 209	6.855 655	21.67948
48	2 304	6.928 203	21.90890
49	2 401	7.000 000	22.13594
50	2 500	7.071 068	22.36068
51	2 601	7.141 428	22.58318
52	2 704	7.211 103	22.80351
53	2 809	7.280 110	23.02173
54	2 916	7.348 469	23.23790
55	3 025	7.416 198	23.45208
56	3 136	7.483 315	23.66432
57	3 249	7.549 834	23.87467
58	3 364	7.615 773	24.08319
59	3 481	7.681 146	24.28992
60	3 600	7.745 967	24.49490
61	3 721	7.810 250	24.69818
62	3 844	7.874 008	24.89980
63	3 969	7.937 254	25.09980
64	4 096	8.000 000	25.29822
65	4 225	8.062 258	25.49510
66	4 356	8.124 038	25.69047
67	4 489	8.185 353	25.88436
68	4 624	8.246 211	26.07681
69	4 761	8.306 624	26.26785
70	4 900	8.366 600	26.45751
71	5 041	8.426 150	26.64583
72	5 184	8.485 281	26.83282
73	5 329	8.544 004	27.01851
74	5 476	8.602 325	27.20294
75	5 625	8.660 254	27.38613
76	5 776	8.717 798	27.56810
77	5 929	8.774 964	27.74887
78	6 084	8.831 761	27.92848
79	6 241	8.888 194	28.10694
80	6 400	8.944 272	28.28427
81	6 561	9.000 000	28.46050
82	6 724	9.055 385	28.63564
83	6 889	9.110 434	28.80972
84	7 056	9.165 151	28.98275
85	7 225	9.219 544	29.15476
86	7 396	9.273 618	29.32576
87	7 569	9.327 379	29.49576
88	7 744	9.380 832	29.66479
89	7 921	9.433 981	29.83287
90	8 100	9.486 833	30.00000
91	8 281	9.539 392	30.16621
92	8 464	9.591 663	30.33150
93	8 649	9.643 651	30.49590
94	8 836	9.695 360	30.65942
95	9 025	9.746 794	30.82207
96	9 216	9.797 959	30.98387
97	9 409	9.848 858	31.14482
98	9 604	9.899 495	31.30495
99	9 801	9.949 874	31.46427
100	10.000	10.00000	31.62278

*Roots of numbers other than those given in this table may be determined from the following relations: *Square Roots:*

$$\sqrt{1000n} = 10\sqrt{10n};\quad \sqrt{100n} = 10\sqrt{n};\quad \sqrt{\frac{n}{10}} = \frac{\sqrt{10n}}{10};\quad \sqrt{\frac{n}{100}} = \frac{\sqrt{n}}{10};\quad \sqrt{\frac{n}{1000}} = \frac{\sqrt{10n}}{100}$$

TABLE III. THE GREEK ALPHABET

Letter	Name	Letter	Name
Α α	alpha	Ν ν	nu
Β β	beta	Ξ ξ	xi
Γ γ	gamma	Ο ο	omicron
Δ δ	delta	Π π	pi
Ε ϵ	epsilon	Ρ ρ	rho
Ζ ζ	zeta	Σ σ	sigma
Η η	eta	Τ τ	tau
Θ θ ϑ	theta	Υ υ	upsilon
Ι ι	iota	Φ ϕ φ	phi
Κ κ	kappa	Χ χ	chi
Λ λ	lambda	Ψ ψ	psi
Μ μ	mu	Ω ω	omega

TABLE IV. PERIODIC TABLE OF THE ELEMENTS

GROUPS

PERIODS	IA	IIA	IIIB		IVB	VB	VIB	VIIB	VIII	VIII	VIII	IB	IIB	IIIA	IVA	VA	VIA	VIIA	O
1	1 H 1.00797																	1 H 1.00797	2 He 4.0026
2	3 Li 6.939	4 Be 9.0122												5 B 10.811	6 C 12.01115	7 N 14.0067	8 O 15.9994	9 F 18.9984	10 Ne 20.183
3	11 Na 22.9898	12 Mg 24.312												13 Al 26.9815	14 Si 28.086	15 P 30.9738	16 S 32.064	17 Cl 35.453	18 Ar 39.948
4	19 K 39.102	20 Ca 40.08	21 Sc 44.956		22 Ti 47.90	23 V 50.942	24 Cr 51.996	25 Mn 54.9380	26 Fe 55.847	27 Co 58.9332	28 Ni 58.71	29 Cu 63.54	30 Zn 65.37	31 Ga 69.72	32 Ge 72.59	33 As 74.9216	34 Se 78.96	35 Br 79.909	36 Kr 83.80
5	37 Rb 85.47	38 Sr 87.62	39 Y 88.905		40 Zr 91.22	41 Nb 92.906	42 Mo 95.94	43 Tc (99)	44 Ru 101.07	45 Rh 102.905	46 Pd 106.4	47 Ag 107.870	48 Cd 112.40	49 In 114.82	50 Sn 118.69	51 Sb 121.75	52 Te 127.60	53 I 126.9044	54 Xe 131.30
6	55 Cs 132.905	56 Ba 137.34	57 La 138.91	58 Ce → 71 Lu	72 Hf 178.49	73 Ta 180.948	74 W 183.85	75 Re 186.2	76 Os 190.2	77 Ir 192.2	78 Pt 195.09	79 Au 196.967	80 Hg 200.59	81 Tl 204.37	82 Pb 207.19	83 Bi 208.980	84 Po (210)	85 At (210)	86 Rn (222)
7	87 Fr (223)	88 Ra (226)	89 Ac (227)	90 Th → 103 Lr	104 Ku (260)	105 Ha (260)	106	107											

LANTHANIDE SERIES	58 Ce 140.12	59 Pr 140.907	60 Nd 144.24	61 Pm (147)	62 Sm 150.35	63 Eu 151.96	64 Gd 157.25	65 Tb 158.924	66 Dy 162.50	67 Ho 164.930	68 Er 167.26	69 Tm 168.934	70 Yb 173.04	71 Lu 174.97
ACTINIDE SERIES	90 Th 232.038	91 Pa (231)	92 U 238.03	93 Np (237)	94 Pu (242)	95 Am (243)	96 Cm (247)	97 Bk (249)	98 Cf (251)	99 Es (254)	100 Fm (253)	101 Md (256)	102 No (254)	103 Lw (257)

Key:

26 — Atomic number (Z) (= number of protons in nucleus)

Fe — Element symbol

55.847 — Atomic mass (in AMU) of the naturally occurring isotopic mixture. 1 AMU = 1.6605×10^{-27} kg = (1/12) × (mass of ^{12}C atom).

For the elements that are naturally radioactive, the numbers in parentheses are mass numbers of the most stable isotope of those elements.

TABLE V. ATOMIC MASSES OF THE NATURALLY OCCURRING ISOTOPIC MIXTURES OF THE ELEMENTS

Element	*Symbol*	*Atomic No.*	*Atomic Mass (AMU)*†	*Element*	*Symbol*	*Atomic No.*	*Atomic Mass (AMU)*†
Actinium	Ac	89	[227]*	Mercury	Hg	80	200.59
Aluminum	Al	13	26.9815	Molybdenum	Mo	42	95.94
Americium	Am	95	[243]*	Neodymium	Nd	60	144.24
Antimony	Sb	51	121.75	Neon	Ne	10	20.183
Argon	Ar	18	39.948	Neptunium	Np	93	[237]*
Arsenic	As	33	74.9216	Nickel	Ni	28	58.71
Astatine	At	85	[210]*	Niobium	Nb	41	92.906
Barium	Ba	56	137.34	Nitrogen	N	7	14.0067
Berkelium	Bk	97	[247]*	Nobelium	No	102	[253]*
Beryllium	Be	4	9.0122	Osmium	Os	76	190.2
Bismuth	Bi	83	208.980	Oxygen	O	8	15.9994
Boron	B	5	10.811	Palladium	Pd	46	106.4
Bromine	Br	35	79.909	Phosphorus	P	15	30.9738
Cadmium	Cd	48	112.40	Platinum	Pt	78	195.09
Calcium	Ca	20	40.08	Plutonium	Pu	94	[242]*
Californium	Cf	98	[249]*	Polonium	Po	84	[210]*
Carbon	C	6	12.01115	Potassium	K	19	39.102
Cerium	Ce	58	140.12	Praseodymium	Pr	59	140.907
Cesium	Cs	55	132.905	Promethium	Pm	61	[145]*
Chlorine	Cl	17	35.453	Protactinium	Pa	91	[231]*
Chromium	Cr	24	51.996	Radium	Ra	88	[226.05]*
Cobalt	Co	27	58.9332	Radon	Rn	86	[222]*
Copper	Cu	29	63.54	Rhenium	Re	75	186.2
Curium	Cm	96	[248]*	Rhodium	Rh	45	102.905
Dysprosium	Dy	66	162.50	Rubidium	Rb	37	85.47
Einsteinium	Es	99	[254]*	Ruthenium	Ru	44	101.07
Erbium	Er	68	167.26	Samarium	Sm	62	150.35
Europium	Eu	63	151.96	Scandium	Sc	21	44.956
Fermium	Fm	100	[253]*	Selenium	Se	34	78.96
Fluorine	F	9	18.9984	Silicon	Si	14	28.086
Francium	Fr	87	[223]*	Silver	Ag	47	107.870
Gadolinium	Gd	64	157.25	Sodium	Na	11	22.9898
Gallium	Ga	31	69.72	Strontium	Sr	38	87.62
Germanium	Ge	32	72.59	Sulfur	S	16	32.064
Gold	Au	79	196.967	Tantalum	Ta	73	180.948
Hafnium	Hf	72	178.49	Technetium	Tc	43	[99]*
Helium	He	2	4.0026	Tellurium	Te	52	127.60
Holmium	Ho	67	164.930	Terbium	Tb	65	158.924
Hydrogen	H	1	1.00797	Thallium	Tl	81	204.37
Indium	In	49	114.82	Thorium	Th	90	232.038
Iodine	I	53	126.9044	Thulium	Tm	69	168.934
Iridium	Ir	77	192.2	Tin	Sn	50	118.69
Iron	Fe	26	55.847	Titanium	Ti	22	47.90
Krypton	Kr	36	83.80	Tungsten	W	74	183.85
Lanthanum	La	57	138.91	Uranium	U	92	238.03
Lawrencium	Lw	103	[259]*	Vanadium	V	23	50.942
Lead	Pb	82	207.19	Xenon	Xe	54	131.30
Lithium	Li	3	6.939	Ytterbium	Yb	70	173.04
Lutetium	Lu	71	174.97	Yttrium	Y	39	88.905
Magnesium	Mg	12	24.312	Zinc	Zn	30	65.37
Manganese	Mn	25	54.9380	Zirconium	Zr	40	91.22
Mendelevium	Md	101	[256]*				

†AMU = 1.6605×10^{-27} kg = $(1/12) \times$ (mass of ^{12}C atom).

*The numbers in brackets are mass numbers of the most stable isotope of those elements. These elements are radioactive.

TABLE VI. USEFUL DATA

Conversion Factors	
1 in	= 2.54 cm (Exactly)
1 mi	= 1.609 km
1 lb	= 0.45359237 kg (Exactly)
1 AMU	= 1.6605×10^{-27} kg
(1 AMU) $\times c^2$	= 931.481 MeV
30 mi/hr	= 44 ft/s
1 eV	= 1.602×10^{-19} J
1 MeV	= 1.602×10^{-13} J
1 Cal	= 4186 J
1 tesla (T)	= 10^4 gauss (G)
Astronomical Data	
1 Light year (L.Y.)	= 9.46×10^{15} m
1 Astronomical unit (A.U.)	=
(Earth-Sun distance)	= 1.50×10^{11} m
Radius of Sun	= 6.96×10^8 m
Earth-moon distance	= 3.84×10^8 m
Radius of Earth	= 6.38×10^6 m
Radius of Moon	= 1.74×10^6 m
Mass of Sun	= 1.99×10^{30} kg
Mass of Earth	= 5.98×10^{24} kg
Mass of Moon	= 7.35×10^{22} kg
Average orbital speed of Earth	= 2.98×10^4 m/s ($\cong$30 km/s)
Physical Constants	
Velocity of light in vacuum	$c = 2.998 \times 10^8$ m/s
Charge of electron	$e = 1.60 \times 10^{-19}$ C
Planck's constant	$h = 6.63 \times 10^{-34}$ J-s
Avogadro's number	$N_0 = 6.02 \times 10^{23}$ mole^{-1}
Electron mass	$m_e = 9.11 \times 10^{-31}$ kg
	$m_e c^2 = 0.511$ MeV
Proton mass	$m_p = 1.6726 \times 10^{-27}$ kg
	= 1.007276 AMU
	= 1836.11 m_e
	$m_p c^2 = 938.26$ MeV
Neutron mass	$m_n = 1.6749 \times 10^{-27}$ kg
	= 1.008665 AMU
	$m_n c^2 = 939.55$ MeV
Gravitational constant	$G = 6.673 \times 10^{-11}$ N-m^2/kg^2

APPENDIX B
ANSWERS TO EXERCISES

Section 1.1

1. 33
2. 3
3. -3
4. 33
5. 270
6. -270
7. -270
8. 270
9. 1.2
10. -1.2
11. -1.2
12. 1.2

Section 1.2

13. $0.333 \cdots$
14. 0.25
15. $0.333 \cdots$
16. 21
17. $1.777 \cdots$
18. 17.5
19. 24
20. $0.628 \cdots$
21. $0.111 \cdots$
22. 3.9
23. 8
24. 2.375

Section 1.3

25. 1 000 000 000
26. 10^7
27. $10^8 = 100\,000\,000$
28. 81
29. 1.728
30. 10^{12}
31. 10^{-5}
32. 10^{-6}
33. 10^3
34. 10^{-7}
35. 10^{-4}
36. 12

Section 1.4

37. 4.07×10^{18} cm
38. 1.67×10^{-24} g
39. 4×10^{3}
40. 1.28
41. 2×10^{6}
42. 2.5×10^{-6}
43. 57.6

Section 1.5

44. 10^{3}
45. 10^{-9}
46. 10^{-6}
47. 10^{7}

Section 1.6

48. 4
49. 9
50. 6
51. 0.5
52. 4.123
53. 0.8
54. 20
55. $9.643 \cdots$

Section 1.7

56. $<$
57. $>, \cong$
58. $\cong$
59. $\propto$

Section 1.8

60. 10^{5}
61. 1.36
62. 1.69
63. 0.62
64. 6.6
65. 50.4
66. 1.23×10^{5}
67. $3.156 \times 10^{7} \cong \pi \times 10^{7}$
68. 2.59×10^{10} cm^{2}
69. 3.79×10^{3} cm^{3}
70. 640
71. 2.2×10^{3} lb

Section 2.2

72. $x = -\frac{1}{2}$
73. $x = -\frac{1}{2}$
74. $x = \frac{21}{4}$
75. $x = 6$
76. $x = \frac{1}{8}(a + b + 7)$
77. $x = 8$
78. $x = \frac{a}{c} - b$
79. $x = \frac{13}{4}$

Section 2.3

80. $x = 15$
81. $x = 10$
82. $x = 7.55$
83. $x = 10$
84. $x = -1, 3$
85. $x = 4, 5$
86. $x = -\frac{19}{6}, \frac{1}{2}$
87. $x = 30$

Section 2.4

88. 3
89. $\frac{15}{6} = 2\frac{1}{2}$
90. 8 g
91. 2 min
92. 550 m

Section 3.1

93. 20.73 m
94. 1.05 m
95. $13° 45'$
96. $216'$
97. (b) 0.5°

Section 3.2

98. 34 600 m^2
99. 125 cm^3

100. 8 cm^3
101. $16\pi = 50.3 \text{ cm}^2$
102. $36\pi = 113.1 \text{ cm}^3$
103. $16\pi = 50.3 \text{ cm}^3$
104. $2.21 \times 10^{19} \text{ m}^3$

Section 3.3

105. 63.2 g
106. 26 kg
107. 11.3 kg
108. $100\pi = 314.16$ kg
109. $\frac{4}{3}\pi = 4.2$ kg
110. $1.4 \times 10^3 \text{ kg/m}^3 = 1.4 \text{ g/cm}^3$ The average density of the sun is only about 40 per cent greater than that of water.

Section 3.4

111. yes; $c^2 = a^2 + b^2$
112. 12 m
113. $10\sqrt{2} = 14.14$ cm
114. $20\sqrt{5} = 44.72$ m

Section 3.5

115. 1.414 ft
116. 0.71
117. 0.71
118. 8 m
119. 0.80
120. 0.600
121. 0.018
122. 0.31
123. 50°
124. 28°

Section 4.1

125. $\$24 \times 10^9$
126. $30 \times 10^6 \text{ km}^3$
127. 0.075
128. 1.5×10^9
129. about 10.8 years

Section 4.2

130. Straight line
131. Square
132. 4 cm
133. $\sqrt{65} = 8.06$ units
134. $\sqrt{61} = 7.81$ units

Section 4.3

135. $x = 8\ t$
136. $x = 8\ t + 6$
137. $x = 8\ t - 10$
138. (1,5)
139. 2.5 s

Section 4.4

140. Displaced by 90°; −1
141. 10 years

Section 5.2

142. $6\vec{x}$ is a displacement of 150 mi directed 40° north of due west.
143. $-1.5\vec{x}$ is a displacement of 37.5 mi directed 40° south of due east.
144. $-4\vec{x}$ is a displacement of 100 mi directed 40° south of due east.
145. $3\vec{x}$ is a displacement of 75 mi directed 40° north of due west.
146. $5000\ \vec{F}$ is a force of 70 000 N directed vertically downward.
147. $-0.5\ \vec{F}$ is a force of 7 N directed vertically upward.
148. $-15\ \vec{F}$ is a force of 210 N directed vertically upward.
149. $13\ \vec{F}$ is a force of 182 N directed vertically downward.

Section 5.3

150. A velocity of 240 m/s directed 45° north of due east.
151. A velocity of 120 m/s directed 45° south of due west.
152. $\vec{R}_F$ is a force of 2.83 N directed at an angle of 45° with respect to the positive X-axis.
153. A velocity of 316 mi/hr directed 18° east of due north.

Section 5.4

154. $F = 500$ N
155. $\theta = 0°$
156. $F_x = 500$ N
157. $F_y = 0$
158. $F = 500$ N
159. $\theta = 90°$
160. $F_x = 0$
161. $F_y = 500$ N
162. $v = 10$ m/s
163. $\theta = 20°$
164. $v_x = 9.4$ m/s
165. $v_y = 3.4$ m/s
166. $F = 20$ N
167. $\theta = 45°$
168. $F_x = 14.2$ N
169. $F_y = 14.2$ N
170. $F = 22.4$ N
171. $\theta = 63°$

INDEX